AF760051

Introduction to the Air Transport System

Milica Kalić
University of Belgrade
Faculty of Transport and Traffic Engineering

Slavica Dožić
University of Belgrade
Faculty of Transport and Traffic Engineering

Danica Babić
University of Belgrade
Faculty of Transport and Traffic Engineering

CRC Press
Taylor & Francis Group
Boca Raton London New York

CRC Press is an imprint of the
Taylor & Francis Group, an **informa** business
A SCIENCE PUBLISHERS BOOK

Cover Credit: Cover illustration courtesy of Teodora Kumanudi

First edition published 2022
by CRC Press
6000 Broken Sound Parkway NW, Suite 300, Boca Raton, FL 33487-2742

and by CRC Press
4 Park Square, Milton Park, Abingdon, Oxon, OX14 4RN

© 2022 Taylor & Francis Group, LLC

CRC Press is an imprint of Taylor & Francis Group, LLC

Reasonable efforts have been made to publish reliable data and information, but the author and publisher cannot assume responsibility for the validity of all materials or the consequences of their use. The authors and publishers have attempted to trace the copyright holders of all material reproduced in this publication and apologize to copyright holders if permission to publish in this form has not been obtained. If any copyright material has not been acknowledged please write and let us know so we may rectify in any future reprint.

Except as permitted under U.S. Copyright Law, no part of this book may be reprinted, reproduced, transmitted, or utilized in any form by any electronic, mechanical, or other means, now known or hereafter invented, including photocopying, microfilming, and recording, or in any information storage or retrieval system, without written permission from the publishers.

For permission to photocopy or use material electronically from this work, access www.copyright.com or contact the Copyright Clearance Center, Inc. (CCC), 222 Rosewood Drive, Danvers, MA 01923, 978-750-8400. For works that are not available on CCC please contact mpkbookspermissions@tandf.co.uk

Trademark notice: Product or corporate names may be trademarks or registered trademarks and are used only for identification and explanation without intent to infringe.

Library of Congress Cataloging-in-Publication Data (applied for)

ISBN: 978-0-367-60921-4 (hbk)
ISBN: 978-0-367-60923-8 (pbk)
ISBN: 978-1-003-10253-3 (ebk)

DOI: 10.1201/9781003102533

Typeset in Palatino
by Radiant Productions

Preface

Air transport operations are expected to double in the next twenty years. This will inevitably create a need to recruit and train thousands of new staff that are necessary to meet this increased demand level. The aviation sector is unique in its dynamics as compared to other service sectors and it encompasses a wide range of jobs and skills. There are many job positions in the air transport sector and most of them require a highly qualified workforce and a significant amount of training. This includes, among others: pilots and air traffic controllers, airline and airport logistics, customer service roles in airline ticketing, check-in, cabin crew and retail, technical engineering jobs, leadership, management and executive roles, etc. One of the issues in the aviation industry is to educate and train human resources in a way that would provide sufficient skills for all the daily operations, as well as for the expected or unexpected technology disruptions/automation.

Due to the projected growth in air traffic, there will be a need to recruit and train thousands of new staff. Although several positions require skill sets specific to aviation, there are many airport, airline, air traffic control, and other aviation stakeholders' occupations that do not belong to specialist aviation jobs. This book is intended for those readers who are not aviation professionals but are involved in the aviation sector and want to learn more about this industry to support their ongoing professional development. It can also be very useful as supporting material for business, management, transportation, and engineering students (undergraduate, master and PhD degree) coming with diverse background knowledge. This book may help them find their way to bridge those gaps by stepping into a world of aviation and embarking on the most epic voyage of all.

The book is divided into seven major parts: introduction to the air transport system, aircraft, airports, air traffic control system, airlines, airline planning and operations, and air cargo. Each area of the air transport system is explained in a manner sufficient to satisfy both those in the industry without any aviation engineering background and the newcomers to the industry with some knowledge of the air transport system.

The opening, Chapter 1, describes the air transport system, its role in the movement of passengers and cargo between countries and provides a historical outline of general aviation, including the past, present, and future trends. This is followed up by fundamentals of aircraft structure and design, the basic principles of flying, and the problems of stability and control of aircraft (Chapter 2). Although every airport is unique in terms of its geographic location, physical layout, demand characteristics, environment, and mix of air traffic, Chapter 3 will provide the basic requirements that must be fulfilled in the design and configuration for every airport. Chapter 3 identifies the roles of principal components of airports and discusses the interrelationships that exist between them; it examines the factors that affect the airport location, orientation, and physical dimensions of runways and other airside components and how these affect the capacity.

Chapter 4 describes how the ATC system works, how it is organized and explains the main terminology in ATC that will assist the readers in achieving an overall understanding of the air traffic control system. The chapter provides explanations that can prepare the readers to understand the procedures used in air traffic control. It also covers the separation rules of aircraft in the ATC system, which refer to the minimum distance between the aircraft operating in controlled airspace and at airports with an operational control tower. Chapter 5 deals with different types of airlines and their business models. Although these categories represent a similar set of business models available in the market, there is some overlapping, which results in numerous types of airlines because their customers have different needs. How each listed business model works is explained and where they see the possibilities of obtaining the revenue they need to earn money and continue with profitable operations.

The main characteristics of demand for air transport services are that it is a derived demand with a range of market segments with differing degrees of price and service quality sensitivity. Chapter 6 explains what the factors influencing the demand for air transport services are, how airlines segment their market, and what the airline supply and its characteristics are. The importance of loyalty programs and the cooperation among airlines is addressed, as well. Types of airline passenger fares are discussed, followed by in-depth coverage of the pricing process and the role of revenue management. This chapter also covers an in-depth discussion of airline operating costs, output determination in the short run and key performance areas with indicators that can be used by the airline management.

Many of the concepts and principles of airline management and operations discussed in previous chapters can be equally applied to air cargo transport. However, there are particular issues and difficulties which arise in the carriage of air cargo and which require separate analysis and treatment, and this is discussed in the final, Chapter 7, of the book.

In a liberalized environment, basic knowledge of engineering, economics, management, and soft skills become of the essence to effectively run a business in aviation. In this context, education and training in the sector will remain the highroad to competitiveness and success in the marketplace. Having this in mind, this book provides a relatively solid and comprehensive educational background for all those involved in air transport and for interested youth who may find this area very attractive and are willing to become a part of the future aviation sector.

November 2021 Milica Kalić, Slavica Dožić and Danica Babić
Belgrade (Serbia)

Acknowledgments

We express our special thanks of gratitude to our colleague, advisor and, above all, our good friend, Dr Milan Janić, for encouraging and supporting us to write this book.

We acknowledge and thank Professor Vojin Tošić, who devoted his time to read and edit our book. His great suggestions and advice helped us refine our ideas and approach throughout the whole book.

We are also deeply thankful to Aleksandar Kukić Kule, Dušan Crnogorac, and Vladimir Trifunović, who patiently helped us bring the images to life so that now they look even better.

A special thanks to Teodora Kumanudi, who made a beautiful artwork for the book cover.

And finally…

We thank our families for putting up with us over the very challenging year it has taken to bring this book to an end.

Contents

List of Acronyms and Abbreviations

A	Altitude
AC	Average Cost
ACI	Airport Council International
ACMI	Aircraft, Crew, Maintenance and Insurance
ADS-B	Automatic Dependent Surveillance-Broadcast
AEA	Association of European Airlines
AIP	Aeronautical Information Publication
ALM	Actual Landing Mass
AOC	Airline Operations Centre
APM	Automated People Mover
ARSR	Air Route Surveillance Radar
ASA	Air Service Agreement
ASK	Available Seat Kilometre
ASM	Airspace Management
ASM	Available Seat Mile
AT/OM	Actual Take-Off Mass
ATC	Air Traffic Control
ATFM	Air Traffic Flow Management
ATM	Air Traffic Management
ATPCO	Airline Tariff Publishing Company
ATS	Air Traffic Services
ATSU	Air Traffic Service Unit
AWB	Air Waybill
AWY	Airway
AZFM	Actual Zero Fuel Mass
BD+C	Building Design and Construction
BRT	Bus Rapid Transit
CAA	Civil Aviation Authority
CAB	Civil Aeronautics Board
CASK	Cost per Available Seat Kilometre
CAT	City Airport Train

CDM	Collaborative Decision Making
CFMU	Central Flow Management Unit
CG	Centre of gravity of the aircraft
CITES	Convention on International Trade in Endangered Species of Wild Fauna and Flora
CORSIA	Carbon Offsetting and Reduction Scheme for International Aviation
CTA	Control Area
CTR	Control Zone
DGR	Dangerous Goods Regulations
DH	Decision Height
DME	Distance Measuring Equipment
DOC	Direct Operating Cost
DSS	Decision Support System
E	Elevation
EASA	European Union Aviation Safety Agency
EAT	Foodstuffs
e-AWB	Electronic Air Waybill
EBITDA	Earnings Before Interest, Taxes, Depreciation and Amortization
ECAC	European Civil Aviation Conference
EMSR	Expected Marginal Seat Revenue
ERAA	European Regions Airline Association
EU	European Union
FAA	Federal Aviation Administration
FAI	Fédération Aéronautique Internationale
FC	Fixed Cost
FFP	Frequent-Flyer Program
FIR	Flight Information Region
FIS	Flight Information Service
FL	Flight Level
FRA	Free route airspace
FSC	Full service carrier
FTK	Freight Tonne Kilometre
FUA	Flexible Use of Airspace
GAT	General Air Traffic
GDP	Gross Domestic Product
GDS	Global Distribution System
GOM	Ground Operations Manual
GPS	Global Positioning System
H	Height
HEA	Heavy cargo
HF	High Frequency

HS	Hub-and-spoke
HUM	Human remains
IAG	International Airlines Group
IATA	International Air Transport Association
ICAO	International Civil Aviation Organization
ID+C	Interior Design and Construction
IFR	Instrument Flight Rules
ILS	Instrument Landing System
IMC	Instrumental Meteorological Conditions
IOC	Indirect Operating Cost
IOSA	IATA Operational Safety Audit
ISIC	International Student Identity Card
ITIC	International Teacher Identity Card
KPA	Key Performance Area
KPI	Key Performance Indicator
LAR	Live Animal Regulations
LCC	Low Cost Carrier
LEED	Leadership in Energy and Environmental Design
LHO	Vaccines, blood and blood plasma, and live human organs
MAC	Mean Aerodynamic Chord
MC	Marginal Cost
MEL	Minimum Equipment List
MEM	Manufactures' Empty Mass
MLAT	Multilateration
MLM	Maximum Design Landing Mass
MRO	Maintenance, Repair, Overhaul
MSL	Mean Sea Level
MT/OM	Maximum Take-Off Mass
MZFM	Maximum Zero Wing Fuel Mass
ND	Neighbourhood Development
NDB	Non-directional (radio) Beacon
NOTOC	Notification to Captain
OAT	Operational Air Traffic
O-D	Origin-Destination
OEM	Operating Empty Mass
OEM	Original Equipment Manufacturers
OIE	Office International des Epizooties
OLM	Operational Landing Mass
OT/OM	Operational Take-Off Mass
P/L	Payload
PAR	Precision Approach Radar
PCR	The Perishable Cargo Regulations
PER	Perishables that are not foodstuff

PN	Total number of passengers
PP	Point-to-point
PPP	Public-Private Partnerships
PSR	Primary Surveillance Radar
QFE	Query Field Elevation
QNE	The International Standard Atmosphere
QNH	Query Nautical Height
RAA	Regional Airline Association
RASK	Revenue per Available Seat Kilometre
REA	Railway Express Agency
RJ	Regional Jet
RPK	Revenue Passenger-Kilometre
RPM	Revenue Passenger-Mile
RTK	Revenue Tonne-Kilometre
RVR	Runway Visual Range
RVSM	Reduced Vertical Separation Minima
SATCOM	Satellite communications datalink
SLI	Shipper's Letter of Instruction
SRE	Surveillance Radar Equipment
SSR	Secondary Surveillance Radar
TA	Transition Altitude
TACT	The Air Cargo Tariff and Rules
TC	Total Cost
TCAS	Traffic Collision Avoidance System
TCR	Temperature Control Regulations
TFC	Total Fuel Capacity
TL	Transition Level
TLY	Transition Layer
TMA	Terminal Control Area
TOC	Total Operating Cost
TP	Turbo-prop
TSA	Temporary Segregated Airspace
UIR	Upper Information Region
UK	United Kingdom
ULD	Unit Load Device
UN	United Nations
UTA	Upper Control Area
VAL	Valuable cargo
VC	Variable Cost
VFR	Visits to Friends and Relatives
VFR	Visual Flight Rules
VHF	Very High Frequency
VMC	Visual Meteorological Conditions

VOR	Very High Frequency Omnidirectional Radio Range
WCO	World Customs Organization
WPX	Worldwide Parcel Express
WTC	Wake Turbulence Category

Chapter 1

Introduction

1.1 Air transport system

The air transport system is a very complex system consisting of several subsystems or components. Its three main subsystems (components) are as follows (Figure 1.1):

- Airports,
- Air traffic control/Air traffic management (ATC/ATM),
- Airlines.

Airports represent the infrastructure in the air transport system, enabling airlines to perform air transport movements (or operations,

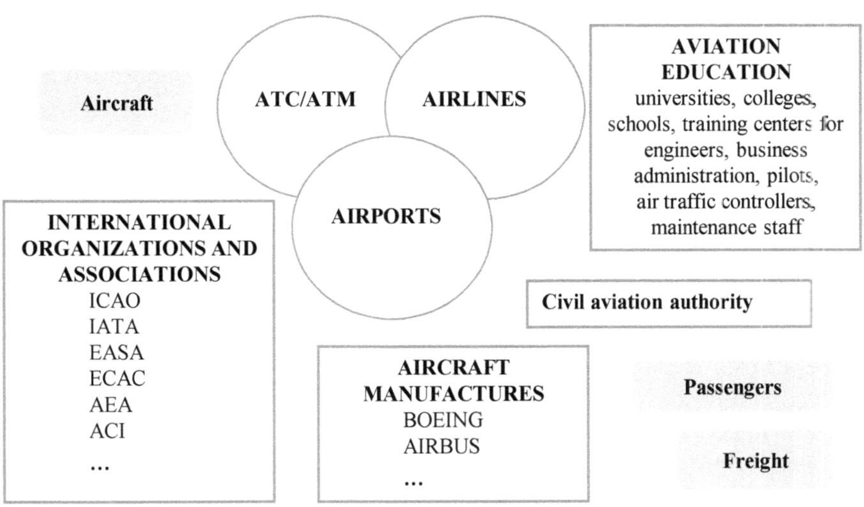

Figure 1.1. Air transport system and its users.

meaning take-offs and landings), as well as to transport passengers and cargo. Airport landside area includes one or several terminals that connect the travel access mode with the air transport mode. Airport air side area covers runways, taxiways, and gates/apron complex, and it is a place where aircraft are serviced by ground handling operators/ground handling department of the airport operator at parking stands, and by ATC controllers during taxiing, take-off, or landing. Airport users are airlines and their passengers, cargo/freight shipments, mail, and aircraft.

ATC/ATM consists of the controlled airspace over particular countries, continents, and oceans (imaginary paths—air corridors), and radio-navigational facilities and equipment located at the ground and in space (satellites) (Janić 2021). Moreover, from the airport ATC tower, controllers monitor all aircraft movements on the ground and control all ground traffic at airport runways and taxiways.

Airlines provide air transport services to passengers, cargo/freight shipments, and mail. Moreover, airlines can transport both passengers and cargo (freight and luggage together in baggage compartment), or only passengers, or only cargo. For airlines, aircraft are 'mobile infrastructure' characterized by different payload capacities and different ranges of flights (Janić 2007). A traditional airline fleet consists of wide-body, narrow-body, and small capacity aircraft. It means that these airlines have a heterogeneous fleet. In contrast, some airlines have a homogeneous fleet consisting of one aircraft type. Depending on their business model, airlines can operate as full service, low-cost, hybrid, charter or regional carriers.

Aircraft have two different functions in the air transport system. First, aircraft is a technical resource for an airline. Second, for airports and ATC/ATMs, aircraft (flights) are users.

All the components are connected and interact among themselves. The relationship between airports and airlines is obvious because these depend on each other. An airport operates due to a decision of airlines to have flights from that particular airport. On other hand, airlines depend on airports in terms of providing safe and secure facilities in the process of passenger, cargo, and aircraft handling. The main common objective of airports and airlines is to fulfill passengers' expectations. Despite this common goal of both airports and airlines, their relationship is very complex because they have different commercial objectives and priorities (Humpreys and Francis 2017). Operational decisions of ATC controllers have a large impact on airline operations and airline costs. Operational decisions refer to route length, or to time spent in holding, which leads to higher fuel consumption and longer flight time.

Airports, airlines, and ATC/ATMs as air transport operators represent the core of the air transport system. All operators balance profitability, safety, security on the one hand, and the users' preferences on the other.

More broadly, the system of air transport encompasses civil aviation authorities, aircraft manufacture, international organizations and associations, aircraft maintenance companies, aviation training centres and schools, educational institutions including universities from the field of aviation, etc.

Each country has a civil (or national) aviation authority (CAA) responsible for the regulation in the field of civil aviation. This authority provides an aircraft register and oversees the approval and regulation of civil aviation. In that sense, CAAs issue safety rules and regulations, certificates of competency for the provision of air navigation services, certificates related to aircraft, airports, aviation agencies, pilot licences, etc. In addition, CAAs perform audits of aviation entities, participate in the work of international civil aviation organizations and institutions and their working bodies, and cooperate with other civil authorities from different countries.

Aircraft manufacturers and suppliers produce aircraft, facilities, and equipment, and sell reliable and safe products to air transport operators (airlines, airports, ATC/ATM). During the process of designing and building aircraft and aircraft parts, manufacturing companies employ high technology. The leading aircraft manufacturers in the world are Boeing and Airbus, followed by Canada's Bombardier Aerospace and Brazil's Embraer.

The most important international organization in the area of civil aviation and public international air law is ICAO. The International Civil Aviation Organization (ICAO) is a specialized agency of the United Nations. It was created in 1944 to promote the safe and orderly development of international civil aviation throughout the world. ICAO is headquartered in Montreal (Canada). ICAO is the primary global civil aviation forum for cooperation between its 193 member countries. "ICAO develops policies and Standards, undertakes compliance audits, performs studies and analyses, provides assistance and builds aviation capacity through many other activities and the cooperation of its Member States and stakeholders."[1] ICAO's global priorities are aviation safety, air navigation capacity and efficiency, security and facilitation, economic development, and environmental protection. ICAO organizes meetings, events, technical cooperation, and offers several different aviation courses and training, etc.

IATA is the International Air Transport Association representing 290 airlines from 120 countries. IATA has headquarters in Montreal (Canada) and the executive offices are located in Geneva (Switzerland).

---

[1] ICAO mission, https://www.icao.int/

IATA members encompass the world's leading passenger and cargo airlines. The goal of this association is to help airlines to operate safely, securely, efficiently, and economically under clearly defined rules. IATA supports and promotes the interests of its members through international recognition and lobbying, assistance related to safety and IOSA standards, global harmonization of security measures and continuous improvement of security processes, organization of conferences, committees, and groups as well as the IATA Annual General Meeting and World Air Transport Summit. IATA offers training through courses in the following areas: air navigation services, airline management, airline operations, and quality, airport planning and operations, cargo and logistics, civil aviation authorities, dangerous goods regulations, environment, fares and ticketing, finance and accounting, ground operations, law and regulations, safety, security, management and leadership, sales and marketing, and travel and tourism. Further, IATA provides many publications covering all aspects of aviation, such as standards, manuals, and guidelines, economics, airlines magazines, annual reviews, etc.

In addition to ICAO and IATA, there are other international aviation organizations, such as Airport Council International (ACI), the European Organisation for the Safety of Air Navigation (EUROCONTROL), European Civil Aviation Conference (ECAC), Association of European Airlines (AEA), European Union Aviation Safety Agency (EASA), etc.

Providing competent human resources is the basis for the progress of an air transport system. Education at all levels plays a key role in developing the competencies of employees in the sector of aviation. It is very important to produce highly proficient engineers, pilots, controllers, dispatchers, mechanics, technicians, flight attendants, and other qualified and competent aviation personnel for all positions in the air transport industry. Besides technical staff in the aviation sector, there are managers who are responsible for overseeing all major operations of the aviation business. Managers differ depending on the aviation sector they work in (airports, airlines, ATC/ATM, manufactures, etc.). The common characteristics that managers have to have are sound judgement, excellent time management, critical thinking, skills, and problem-solving ability. Investments in human resources, in the form of education and training, have crucial importance in sustaining the aviation industry's growth as well as safe and secure air transport operations. There are many institutions offering programs in air transport, such as universities, colleges, schools, training centres, etc. Furthermore, air transport courses are often part of business administration (BA), tourism programs, and general transportation programs.

Using high technology, advanced facilities on the ground and in aircraft, and applying strict rules and procedures lead to well performed air transport services. Well educated and trained employees with licences in air transport sub-systems provide the implementation of all prescribed rules and procedures and ensure the required level of safety and security as well as effectiveness and efficiency. The effectiveness and efficiency are related to providing the required level of service quality and prices on the one hand, and covering operators' costs, on other hand (Janić 2013).

In comparison with other transport modes, air transport is extremely dynamic, expensive, and highly vulnerable to internal or external disruptive events. Thus, it is very important to carefully plan and manage expensive technical and human resources to achieve the desired results in terms of the volume of transport and the level of services provided (Janić 2000). The air transport system should be designed in such a way to enable safe and secure execution of all segments of an air journey (passenger and aircraft handling, and flight operations). A dynamic and complex air transport system is based on cooperation between airports, airlines, ATC/ATM, and other aviation entities, their employees, and on the compatibility of technical systems to consistently enforce the strict rules and procedures. Safety is the air industry's number one priority.[2] During the period from 2008 to 2018, in scheduled commercial air transport, the average number of fatalities in air crashes was 463. In 2017, scheduled commercial air transport accidents resulted in 50 fatalities and that was the safest year ever on the record of aviation.[3] The total number of passengers carried on scheduled services in 2017 was 4.1 billion.[4] According to statistics of the commercial aviation industry, travelling by plane is very safe.

The air transport system is characterized by a high level of cooperation. To provide a seamless and efficient travel experience to air passengers, Collaborative Decision Making (CDM) has been developed and implemented at many airports. CDM provides sharing information among partners and working together more efficiently and transparently in real time. CDM partners are airports, airlines, ATC/ATM, ground handlers, de-icing companies, the network manager, and support services such as police, customs, and immigration. Moreover, the implementation of operational procedures and automated processes creates benefits for the entire air transport system. The common goal is to improve the overall performance of the air transport system in a sense of reducing airport and en route delays and optimizing airport operations.

---

[2] https://www.iata.org/en/programs/safety/

[3] https://www.icao.int/safety/Documents/ICAO_SR_2019_29082019.pdf

[4] https://www.icao.int/annual-report-2017/Pages/the-world-of-air-transport-in-2017.aspx

1.2 Definition of aircraft

According to ICAO Annex 1, an aircraft is any machine that can derive support in the atmosphere from the reactions of the air other than the reactions of the air against the earth's surface. There are several different classifications of aircraft, the following being only one of them. Generally, aircraft can be divided into two broad categories: aerostats and aerodynes.

- Aerostats gain lift using a buoyant gas, such as helium or hydrogen, therefore they are lighter than air. Balloons and airships belong to aerostats.
- Aerodynes are heavier than air aircraft, deriving their lift from motion. Conventional fixed-wing aircraft (airplanes) and helicopters, gyrocopters, and autogyros are aerodynes.

1.3 Brief history of aviation

The first invention and the first event that marks the beginning of aviation development was the balloon that soared into the air. This happened in 1783 in France, thanks to the Montgolfier brothers (Joseph-Michel Montgolfier (1740–1810) and Jacques-Étienne Montgolfier (1745–1799)). The first people flew in a balloon that same year, and two years later, Jean-Pierre (François) Blanchard (1753–1809) was the first person who made a flight over the English Channel in a balloon, with one companion.

This great event was a stimulus to many inventors, researchers, and scientists who dealt with aerodynamics and aircraft construction. Although they had clear and different visions of the possibilities for flying, it took many years to reach a certain level of development of science and technology.

Henri Giffard (1825–1882) solved the problem of aerostat control when he installed a steam engine on an elongated balloon 44 meters long. This airship flew at a speed of 9 km/h with a pilot who changed the course during the flight and made one 360-degree turn. This French scientist made the first successful flight by air in 1852 (Petit 1967, Zorić 1977).

Sir George Cayley (1773–1857) was an English scientist and researcher in the field of aviation and the first person who understood the basic principles of aircraft propulsion. In 1809, he gave the first clear picture of an airplane. He was the first to notice the connection and define the forces of lift, drag, weight, and thrust.

The most important work of Otto Lilienthal (1848–1896) was to determine the relationship between lift force on one side, and the curvature of the wing, and the angle of attack of the wing in relation to the air current, on the other side. He drew his measurements as a polar diagram that is still used today. Germany's Otto Lilienthal made several gliders, similar to today's dragons, which he could control and perform real flights (Nejhum 2006).

With the advent of gliders, it became possible to fly with the help of wings. However, the duration of such a flight was short. To fly a longer distance, an engine was necessary. The steam engines were either heavy or too weak. It was necessary to wait for the appearance of suitable engines for motor flying to come to life (Nejhum 2006).

At the end of the 18th and during the 19th century, predominantly aerostats developed, but at the same time preconditions were created and knowledge necessary for the realization of airplane flight was accumulated (Figure 1.2).

At the beginning of the 20th century, the first flight by plane was made. After the first flight, the successes in flying were followed with airplanes that did not even remotely look like today's airplanes. These were aircraft without a cabin or with open cabins. On December 17, 1903, the first flight of an airplane designed by the Americans, the Wright brothers, Orville (1871–1948) and Wilbur (1867–1912), was officially recorded.

On this day, the Wright brothers performed three flights, but only the third attempt was successful. In the third attempt, they made a flight that lasted 59 seconds and flew a distance of 260 meters. The flight was performed in front of five witnesses, one of whom took photographs (Petit 1967). This most significant event in aviation took place in Kitty

XVIII century	XIX century		
	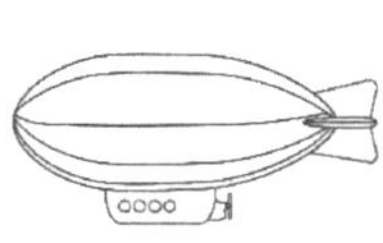	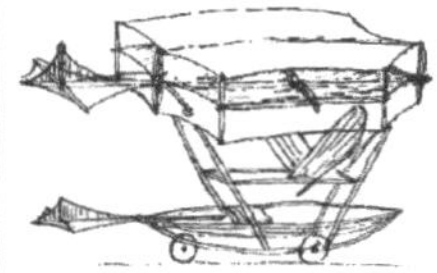	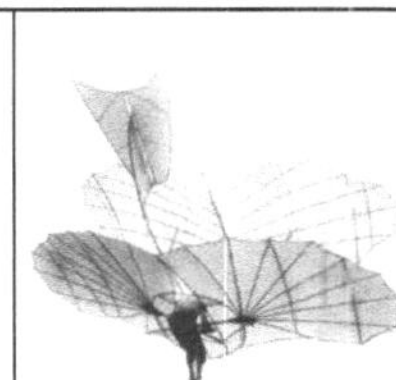
BALLOON	AIRSHIP	GLIDER WITH A BASKET FOR PASSENGER	HANG GLIDER
The Montgolfier brothers (Joseph and Étienne)	Henri Giffard	Sir George Cayley	Otto Lilienthal
France	France	United Kingdom	Germany
Year 1783	Year 1852	Year 1849	Year 1891

Figure 1.2. Development of aviation in the 18th and 19th centuries.

Hawk (North Carolina, US), on the North Atlantic coast. This event and the date when it happened are considered the beginning of the development of aviation based on dynamic aircraft.

In the earliest days of aviation development, there was little need for an organized system of air traffic control. Almost all of the air traffic in the world was conducted in daylight with clear flying conditions, meaning that pilots could fly only in conditions that would permit them to see other aircraft (Nolan 2011).

In the period from 1903 to 1920, airplanes were improved, and thus became a means of transporting passengers, goods, and mail. From day to day, from month to month, from year to year, planes flew at higher speeds, reached greater distances, and flew at higher altitudes. Table 1.1 shows the records reached in 1909 and 1910.

> *Interesting fact—THE FIRST WOMAN WITH A PILOT'S LICENCE*
>
> On March 8, 1910, Women's Day was declared in Copenhagen. On the same day, the first woman received a pilot's license in France and around the world. It was Baroness Raymond de Laroche (1882–1919). The World Air Sports Federation (*Fédération Aéronautique Internationale* – FAI) issued licence no. 36, while the Aero-Club of France issued her the diploma of pilot aviator for women number 1 (Tonković 2017).

Initially, airplanes were used to transport mail. Thus, in 1911, in England, and in 1912 in Germany, the plane was used to transport mail. In these and the following years, mail was transported by plane all over the world. In 1914, World War I began, and then the plane began to be used in war operations.

> *Interesting fact—ROLAND GARROS*
>
> When we say Roland Garros, the first and the only thing we think of is the tennis tournament (one of the four most important in the world) which takes place in Paris on the courts of Roland Garros. But, do you know who Roland Garros was? He was the state cycling champion (in 1906) at the age of 18, and four years later he began his aviation career at the Louis Bleriot pilot school near Paris. He participated in air races until World War I. He was the first pilot to successfully fly the Mediterranean, from the south of France to Bizerte in Tunisia, and that was in 1913. He was only 30 years old when he died in the final phase of World War I (1918). Today, the international airport on his native island of Reunion is named after him, as well as a sports centre with a complex of tennis courts in Paris (since 1920).

Table 1.1. Records of speed, range and altitude in 1909 and 1910.

Date	Aircraft	Motor	Place	Record holder	Speed record
23.08.1909	Curtis	Curtis	Reims, France	G. Curtis	69.82 km/h
28.08.1909	Bleriot	E.N.V.	Reims, France	L. Bleriot	76.99 km/h
23.04.1910	Antoinette	Antoinette	Nice, France	H. Latham	77.58 km/h
10.07.1910	Bleriot	Gnome	Reims, France	L. Moran	106.51 km/h
29.10.1910	Bleriot	Gnome	New York, USA	A. Leblanc	109.76 km/h
Date	**Aircraft**	**Motor**	**Place**	**Record holder**	**Range record**
25.08.1909	Voisin	Gnome	Bétheny, France	L. Paulhan	134.0 km
04.11.1909	M. Farman	Gnome	Mourmelon-le-Grand, France	H. Farman	234.2 km
20.07.1910	Bleriot	Gnome	Mourmelon-le-Grand, France	J. Olieslagers	392.7 km
11.12.1910	Bleriot	Gnome	Pau, France	G. Legagneux	515.9 km
30.12.1910	M. Farman	Reno	Buc, France	M. Tabuteau	584.7 km
Date	**Aircraft**	**Motor**	**Place**	**Record holder**	**Altitude record**
18.10.1909	Wright	Wright	Paris, France	C. de Lambert	300 m
01.12.1910	Antoinette	Antoinette	Châlons, France	H. Latham	453 m

Source: Petit 1967.

After the end of World War I, airplanes began to be used to transport passengers, so the first airlines were established in this period. In 1919, the first scheduled air service (London-Paris) was established by the British airline named Aircraft Transport and Travel Limited (this company was founded during World War I). In 1919, the Dutch airline KLM (Koninklijke Luchtvaart Maatschappij NV) was founded. After that, other airlines were founded (the Colombian airline Avianca (1919), the Australian Qantas (1920), the Russian Aeroflot and the Czech ČSA (Československé státní aerolinie) today's Czech Airlines (1923)).

Bearing in mind that aircraft were more often used as a means of commercial transport, the need to regulate this area of aviation by legal acts arose. In 1919, the Paris Convention (formally, the Convention Relating to the Regulation of Aerial Navigation) established the basic principles of aviation law.

It is interesting that at the same time as airplanes, airships were used to transport passengers. In 1909, the first air passenger company was founded, and it had a fleet of five airships. It was the German company DELAG (Deutsche Luftschiffahrts-Aktiengesellschaft), which transported more than 37 thousand passengers in less than three decades. Even in the period after World War I, airships were used to transport passengers. Cabins were extremely nicely furnished, resembling saloons, and passengers were served by cabin staff during the flight. It is interesting to note that the first flight attendant in the world was a certain Heinrich Kubis (1888–1979), who started working in 1912 and who survived a major accident in 1937 when an airship caught fire. The last large airship was built in 1935. However, low speed, difficult manoeuvring, dependence on the wind, and the danger of fire, as well as a large number of accidents led, just before World War II, to the withdrawal of the airship from air traffic activities (Zorić 1977).

In the 1920s and 1930s, records were broken in terms of range (distance flown), flight speed, and altitude at which the aircraft flew (balloon, seaplane, plane, glider). In the period from 1920 to 1937, the records of the achieved distance, altitude, and speed increased year after year. The highest speed of 709 km/h was achieved by an Italian test pilot Francesco Angelo (1902–1942), in 1934, with a seaplane.

Some other unusually brave pilots—celestial aces inscribed on the pages of aviation history should also be mentioned. They are: Charles Lindbergh (1902–1974), Charles Kingsford Smith (1897–1935), Amelia Earhart (1897–1937) and Jean Batten (1909–1982). For example, in 1927, Charles Lindbergh flew across the Atlantic without a stopover on the New York-Paris route and gained incredible popularity around the world. It was a non-stop flight that lasted 33 hours and 30 minutes, he flew a distance of 6,200 km (or 5809 km, different sources give different values), alone in the cabin of the *Spirit of St. Louis* plane. He flew at altitudes of 1,500 to 3,000 meters (Petit 1967). Lindbergh's success caused a great of progress in civil aviation both in the United States and in the world. Airlines were established, and passenger traffic became more and more important from day to day. But it should be noted that eight years before Lindbergh, Sir John Alcock (1892–1919) and Sir Arthur Whitten Brown (1886–1948) were actually the first to cross the Atlantic, from Newfoundland in Canada to County Galway in Ireland (a distance of just over 3,000 km, which they flew in 16 hours).

In 1928, Amelia Earhart (1897–1937) became famous all over the world, because she was the first woman to cross the Atlantic Ocean. She was on the plane with Wilmer Stultz (1900–1929) piloting. She wanted to cross the Atlantic on her own, and succeeded in doing so in 1932. Charles Kingsford Smith (1897–1935) was an Australian pilot who flew the Pacific Ocean in

1928 from the United States to Australia (Oakland (California) - Hawaii - Fiji - Brisbane). New Zealander Jean Batten (1909–1982), after crossing the South Atlantic in 1935, was the first to make a solo flight from England to New Zealand, and this was in 1936. Unfortunately, many celestial aces of that time, after a number of realized flights and records were achieved, lost their lives in plane crashes.

The first air traffic control was established in the early 1930s to provide conditions for safe flying. In the beginning, air traffic controllers gave instructions to pilots during the take-off and landing by standing in a prominent location on the airfield and waving green and red flags. The flags were later replaced with green and red-light signals (light guns). A light gun is a device that allows the controller to direct a narrow beam of high-intensity coloured light to a specific aircraft (Nolan 2011). In the mid-1930, the first radio link was introduced between the pilot and the controller that permitted direct voice communication with pilots over a certain distance. This radio link allowed controllers to issue traffic instructions, weather information, and airport landing conditions to pilots of radio-equipped aircraft in all weather conditions, during the day or night.

Further development of air transport caused the establishment of airways and navigation aids as well as the emergence of air traffic control units that had the responsibility to monitor and separate aircraft on these airways. In the second half of the 1930s, air traffic rules were established (Instrument Flight Rules—IFR) to be followed by pilots when flying on airways in instrument meteorological conditions (Nolan 2011). Traffic regulation was procedural based on radio communication with pilots. In this way, controllers received position reports from pilots, which they plotted on airway maps and updated the aircrafts' information on note cards known as strips.

It is considered that the plane was most developed and most improved during the wars (World War I (1914–1918), World War II (1939–1945)). This is not surprising, because, in those times, funds were not saved, but progress was sought. That is why many sciences that were directly applied in aviation, such as mathematics, physics, the strength of materials, meteorology, metallurgy, and medicine, advanced thanks to aviation (Zorić 1977).

The Convention on International Civil Aviation (the Chicago Convention) was held in Chicago in 1944, and it was attended by representatives of 52 states. Three separate agreements were made at the convention: the establishment of ICAO (International Civil Aviation Organization), the International Air Services Transit (Two Freedoms), the Agreement (Transit Agreement), and the International Air Transport

Agreement. In April 1945, the International Air Transport Association (IATA) was founded by 31 representatives of air carriers in Havana (Cuba).

The end of World War II was followed by a period of aircraft improvement in terms of all aspects, both technical and economic. Table 1.2 shows the characteristics of representative aircraft in the period from 1926 to 1975.

Regarding the ATC, after World War II new navigation equipment was introduced: Very High Frequency Omnidirectional Radio Range (VOR) and Distance Measuring Equipment (DME). Also, traffic surveillance radar equipment was installed for the first time at busy airports and at air traffic control centres (Surveillance Radar Equipment—SRE) to assist controllers to separate aircraft. To improve the capabilities of aircraft attempting to land in poor weather conditions, the Precision Approach Radar (PAR) and Instrument Landing System (ILS) were installed (Nolan 2011).

The 1950s and 1960s are known as the golden age of travel. Regarding the time horizon from 1960 to the introduction of the Deregulation Actin the US in 1978, it can be seen that this period encompasses the golden age of flying. Looking back at air travel in the 1960s and 1970s, one can notice that travelling by aircraft was a special experience for passengers. Namely, comfort in this period was much higher in comparison to today's aircraft. The passenger cabin was designed with larger legroom and wider aisles, which provided a more comfortable journey. In-flight service was also on a higher level including meals and beverages, and attractive flight

Table 1.2. Representative aircraft in the period from 1926 to 1975.

Manufacturer	Designation	Introduced	Engine power (hp)	Speed (km/h)	Range (km)	Capacity (seats)
Ford	Trimotor	1926	3 × 300	184	800	11
Boeing	B 400 D	1929	1 × 525	176	856	4
Douglas	DC 3	1935	2 × 1,200	272	2,200	21–74
Boeing	Clipper	1939	4 × 1,500	264	5,600	74–89
Douglas	DC 6B	1947	4 × 2,100	456	4,100	32–101
Vickers	Viscount	1948	4 × 1,540	505	2,300	53 max
Martin	404	1950	2 × 2,400	450	1,700	40 max
Boeing	B 707.320	1959	4 × 8,000	960	12,200	147–200
Douglas	DC 9	1965	2 × 5,300	840	1,750	56–115
Boeing	B 747	1969	4 × 18,200	985	12,800	363–490
BAC/Ltd Sud-Aviation	Concorde	1976	4 × 14,400	2,300	6,400	136–146

Note: table is partially taken from Zorić 1977.

attendants, and the impression was that travel was considered a luxurious experience. However, another characteristic of flying in the golden age was the fact that it was much more expensive in comparison with the present day.

This period is also recognised as the jet era due to the advent of jet aircraft. Namely, in this period many aircraft made their first flight and were introduced into the service. Some of them were: Boeing 727 trijet (the first flight made in 1963, introduced in 1964), DC-9 twinjet (the first flight made in 1965, introduced in 1965), Boeing 737 (the first flight made in 1967, introduced in 1968), Boeing 747 jumbo jet (the first flight made in 1969, introduced in 1970), Concorde (the first supersonic commercial aircraft—the first flight made in 1969, introduced in 1976), and Airbus A300 (the first flight made in 1972, introduced in 1974), Figure 1.3 (Boeing 2021).

In 1956, the first air route surveillance radar (ARSR) was used in air traffic control centres (Nolan 2011). Soon after, development began on a secondary radar system that used transmitters/receivers and transponders in each aircraft to display the aircraft identification, position, and altitude on the controller's radar screen.

In the 1960s and 1970s, air traffic control made significant progress in terms of transition to the radar system of traffic regulation, the introduction of the Inertial Navigation System, and the beginning of the automation era. Computer systems were beginning to transform air traffic control, allowing for the first time data from the flight plan to be correlated with readings from the radar and transponder, producing alphanumeric screen readouts of data on the plane's position, speed, and altitude (Nolan 2011).

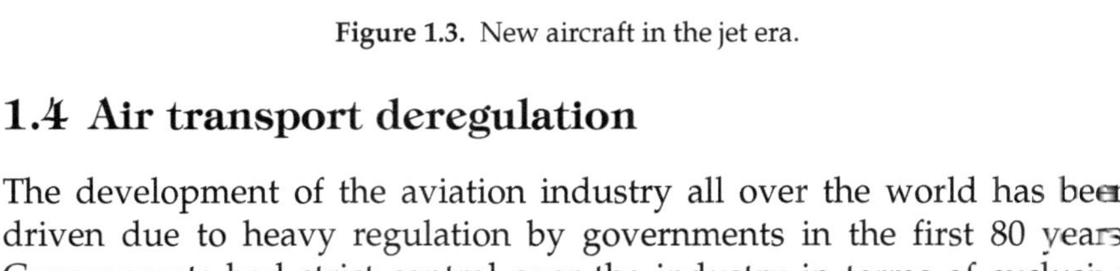

Figure 1.3. New aircraft in the jet era.

1.4 Air transport deregulation

The development of the aviation industry all over the world has been driven due to heavy regulation by governments in the first 80 years. Governments had strict control over the industry in terms of exclusive ownership control of airlines, markets served, city-pair frequency, prices, and market entry of new airlines. In the US, regulatory control was assigned to the Civil Aeronautics Authority (the precursor to the Federal Aviation

Administration– FAA) and the Civil Aeronautics Board (CAB) which was responsible predominantly for economic regulation. CAB regulated airline routes, prices, mergers, acquisitions, and market entry (Fischer 2018). Due to the price control, the airlines competed with each other mostly on the basis of the service quality (better food, more comfortable seats, etc.).

At the same time in Europe, air transport developed under the control of national authorities, which means that routes were operated by national airlines, and airports were publicly owned. The whole industry relied on bilateral air service agreements (often called ASAs). A bilateral ASA is concluded between two contracting countries and it liberalizes commercial civil aviation services between these countries. Based on this agreement the designated airlines of those countries (frequently allowing only one airline from each country to operate the services) were allowed to operate on a limited number of specified routes and airports, with regulated frequencies, using the agreed aircraft types, and charging the specified prices (OECD 2014). This means that the capacity on the majority of routes was artificially constrained, prices were high and market entry by non-flag airlines was impossible.

Deregulation of the air transport industry started in the United States in the mid-1970s. It did not happen instantly, but more as a process measured in years of the gradual abolition of government regulation (economic, legal, administrative, and political). It significantly changed the market environment and the nature of the competition between airlines. Although the deregulation reduced the government's role particularly in the economic area, the government still has significant regulatory authority[5] (Aviation Industry Regulation).

Unlike the market deregulation process in the US, the challenge within Europe was to associate and harmonize several distinct national markets into a single market, previously interlinked by numerous bilateral ASAs. As a result, the process of liberalization in Europe[6] was gradually implemented through four packages (Button 2001):

1. In 1987, the first package of liberalisation measures was applied through the development of two international agreements which permitted partial capacity and tariff liberalization. Airlines were given flexibility for cooperation within the limits of the existing ASAs. For example, airlines were free to modify prices unless both countries disapproved.

[5] For example, the governments are responsible for the regulation of aviation safety and security, protection of competition and the environmental interest, etc.

[6] Liberalization of the air transport market in Europe covers the full market liberalisation within the EU with extension to cover Iceland, Norway and Liechtenstein.

2. In 1990, the second package of liberalisation measures was applied and allowed all European airlines to carry passengers to and from their home countries to other EU member states (3rd and 4th freedoms of the air). Also, the 5th freedom of the air, i.e. the right in respect of scheduled international flight with a stop-over in a third country and the right to pick-up and drop-off passengers during the stopover, was allowed to a greater extent. Price and capacity restrictions were further abolished.
3. In 1993, the third package of liberalisation measures came to take effect, including the common licencing of airlines and freedom of access to the market. All airlines holding a community license were allowed to serve any international route within the EU. With these measures, significant barriers to entry into scheduled services were removed, along with setting common rules governing safety and financial requirements. This gave the right to EU airlines to fly between member states without restrictions and within member states (other than their own) subject to some controls on prices and capacity.
4. From 1997, as part of the third liberalisation package, full cabotage was permitted to all airlines holding a community license, i.e., the right to operate domestic routes within the EU. Moreover, the prices were generally unregulated. With this final package, all remaining commercial restrictions for European airlines operating within the EU were removed, thus setting up the European Single Aviation Market.

The remaining international air transport is still primarily governed by the so-called Chicago regime of restrictive bilateral ASAs. This means that all aspects of air transport commercial relations between nations in the rest of the world still have to be negotiated through bilateral ASAs. Bilateral ASAs expanded over time into multilateral air service agreements that involved more than two contracting states. These agreements have in some cases led to the signing of another form of agreement known as the open skies agreement[7] (Burghouwt and De Wit 2005).

[7] Open skies agreements are policies designed to create an international, free-market environment for the airline industry. Open skies agreements eliminate government interference in the commercial decisions of airlines about routes, capacity, and pricing, allowing them to provide more affordable, convenient, and efficient air service for consumers (Button 2001). Open skies agreements are established to replace the standard ASA giving the two countries' airlines the right to fly where and when they want between the two nations (and beyond) and to set their own prices accordingly (OECD 2014). The EU-US Open Sky Agreement was signed in April 2007 and went into effect in March 2008.

1.4.1 Air transport market after deregulation

The period after the introduction of the Airline Deregulation Act in 1978 was marked by several phases of expansion and cutback, with significant effects on the industry's structure, profitability, patterns of service, and average prices. In the first phase, the influx of new entrants happened in many markets that had been previously protected by the CAB and new airlines were formed. Most of the new airlines followed the low-cost business model (Peterson 2018). Airlines were given complete freedom to determine their own strategies to meet the demand, and become competitive both in terms of prices and quality of service (frequency of flights, the markets they serve, etc.). As a result of competition from new airlines, the already established airlines were forced to reduce their costs, i.e., to enhance their survival and profitability in the competitive market.

With the given freedom and flexibility in decision-making, airlines restructured their networks into a hub-and-spoke route structure, which feeds passengers by spoke routes into a major airport (hub) from which passengers can continue their journey by taking the connecting flights to their final destination. Hub-and-spoke network strategy provided cost savings from more efficient use of labour and fuel associated with larger aircraft, on the one side, and the ability to offer frequent direct and indirect single-airline services between many city-pair markets, on the other side (Morrison and Winston 1986).

In the early ages of deregulation, a new pricing scheme concept, known as yield management, was adopted by major airlines to maximize revenues gained from ticket sales on each flight. This concept was first developed by Boeing to charge passenger seats on the same flight at different prices by using restrictions to segment passengers (Peterson 2018). The greatest progress was made in terms of technology. Very soon after the deregulation, airlines were aligning their fleets with deregulated route structures and markets by using new fuel-efficient medium-sized jets (e.g., Boeing 737-300s, 767, 757 Airbus 321s) (Morrison and Winston 1986). This caused a significant drop in the cost of air transport during the first decade of the deregulation, the average prices declined, which further stimulated a rapid increase in demand for this type of service (Transportation Research Board 1999, US Department of Transportation 2000).

Technology has continued to reduce transport costs, but in the late 1980s, airlines started to seek alternative ways for further reduction of costs. The biggest impact on cost reduction was related to the change of strategy in terms of labour relations. Airlines started to outsource the services, reduced the number of employees and the employee wages to stay competitive (Peterson 2018).

During this period another trend came onto the scene. A new model of feeder service emerged, the major airlines started subcontracting the services on low-density short-haul and medium-haul routes to regional airlines. This was driven by the advent of regional jets with fewer than 75 seats in the cabin that allowed better adjustment of supply with the passenger demand on shorter routes (Peterson 2018).

Commercial airlines have long been protected from competition, both by other airlines and by other modes of transport. It was expected that such a large increase in demand in the period after the deregulation would provide companies with certain financial stability, but this was not the case in air transport. Several airlines managed to remain profitable every year, but looking at the entire branch of the industry, it is only marginally profitable (Dempsey 2008). Many of them failed to adapt to the new market conditions, and instead, they created large losses and eventually withdrew from the market (e.g., Trans World Airlines, Pan American World Airways, Continental Airlines, etc.). In the period between 1983 to 1993, the full-service airlines in the US were able to grow through mergers/acquisitions and bankruptcies leading to increased concentration and market share of the few dominant players, e.g., American, Continental, Delta, Northwest, United, and US Air (Goetz and Vowles 2009). This trend was terminated in the mid-late 1990s when another wave of expansion occurred and new entrants such as Frontier, JetBlue, Kiwi, Midway, etc., stepped in to the market. The dominant position of larger airlines was undermined and they experienced a decline in their market share (Goetz and Vowles 2009).

In the EU, market liberalisation has significantly changed the market environment and the nature of competition between airlines (Button 2001). EU rules ensure fair competition conditions between all airlines and thereby safeguard connectivity to and from the Union (Regulation (EC) No 868/2004). Airlines obtained unlimited freedom to choose their routes, capacity, schedules, and prices. Same as in the US, full-service airlines adopted hub-and-spoke networks to benefit from cost- and demand-side efficiencies, as well as to protect themselves from the new entries (Button 2001).

The availability of airport capacity and airport services were major barriers to entry on the EU market, and thus many new regulations were adopted. For example, Regulation (EEC) No 95/93 provides fair access to airports and airport services and ensures that at congested airports 'slots' shall be allocated to airlines in an equitable, non-discriminatory, and transparent way by an independent 'slot coordinator'. Also, Directive 96/67/EC ensures competition at the market for ground handling services (i.e., the services provided to airlines at airports, such as passenger and baggage handling, fuelling and cleaning of aircraft, etc.). EU competition rules also include rules on state aid, which were first applied to EU airlines in the 1990s, but are now applied to airports, as well, forcing them to behave as 'commercial undertakings' and hence, not allowing them to favour certain airlines (Burghouwt et al. 2015). Most of the major airports have become shopping malls in addition to being the transport hubs (Peterson 2018).

However, deregulation did not have a positive effect in terms of airport construction and expansion. There are few new airports that have been constructed and only a few runways have been added at congested airports. Airport construction faces significant non-political barriers, such as urban development objection and environmental noise and emissions considerations.

The reform led to greater competition both on EU domestic routes and on international routes within the Union (Button 2001). This resulted in a rise in the number of new routes served, the market definition of ticket prices, the entry of new airlines into the market, and greater competitiveness among airlines.

US domestic airline deregulation was also very successful in changing the conditions in the international aviation regulatory environment. There have been a large number of liberalized open skies agreements signed with other states. This initiative led to the reduction of international prices, an increase in the number of international flights and passengers, and an increase in the number of cities with international service (Goetz and Vowles 2009). Most importantly, open skies agreements allowed airlines of the signatory countries to extend their routes, merge with other airlines and establish alliances with foreign partners (Peterson 2018).

Bilateral ASAs were insurmountable barriers to entry on international routes since airlines were required to follow the route rules established in ASAs, which sometimes prevented them from serving points they preferred. To overcome these barriers airlines started to cooperate from the earliest days of the deregulation. In the beginning, this cooperation was focused on interline ticketing and baggage agreements. Codesharing became common in the early 1990s, OECD 2014. A major turning point for airline cooperation happened in 1992 by the signing of the open skies agreement between the US and the Netherlands when the launch of the modern era of airline global alliances began. In network planning, international airlines design a route structure that includes routes between their home country and another, and often, routes beyond that country to a third country (OECD 2014). In alliance with another airline, an airline can expand its network and provide customers with many more itinerary combinations than it could on its own (Cento 2009).

Air traffic control (ATC) has also undergone changes after the deregulation. Many countries privatized their air traffic control service providers (e.g., Canada) to implement cost-based structures and to apply pricing mechanisms to allocate the scarce resource of air traffic capacity with the aim of reducing congestion, as well as other resources more efficiently. The US air traffic control system is still designed and managed under the control of the FAA. In Europe, the European Commission attempts to reform ATC that aims to reduce increasing delays to consumers by taking a central network management approach under EUROCONFTROL and implementing capacity contracts with national ATC providers (Button 2001). This initiative started in 2001 under the Single European Sky legislative framework, but the overall process of reforming the European airspace has been very slow. The main challenge in ATC structural reform is the regulation of safety. However, the proponents of this idea argue that keeping aircraft safely separated is a purely operational process that can be performed effectively by a non-governmental entity as long as it is subject to oversight by safety regulators (Robyn 2018).

The period after 2000 has been marked by a large number of negative events (e.g., the terrorist attacks on September 11, 2001, the SARS pandemic, the invasions of Afghanistan and Iraq, economic recession, and rising fuel costs) that hit the aviation industry very hard. These events were followed by prolonged periods of low demand that led to the loss of nearly USD35 billion from 2001 to 2005, US Air Transport Association 2006. Many large airlines such as Delta, Northwest, United, and US Airways (twice) filed for bankruptcy protection since 2002, while some of them were forced to look for other ways of surviving (e.g., TWA was acquired by American in 2001, US Airways merged with America West in 2005). In the EU some of the full service airlines fully disappeared from the market (e.g., Malev and Slovak Airlines), while others with financial problems were established as new national airlines under a new name (e.g., Sabena, Swiss Air, Balkan Air, Olympic Airways, etc.) (Burghouwt and De Wit 2015). The period after 2000 will certainly be remembered by the cross-border mergers/acquisitions that occurred within the EU and the formation of the 'big three', i.e., Air France – KLM, Lufthansa Group, and IAG.

Consolidation did not circumvent leisure airlines, either. This segment of the airline industry has also undergone significant changes after 1993 and the third package of liberalization. The consolidation process among the tour operators and the highly competitive pressure coming from LCCs started the consolidation process in the holiday charter market (e.g., TUI Travel and Thomas Cook Group) (Burghouwt et al. 2015).

In parallel, the low-cost segment was profitable during this period, and not as badly affected as the major airlines. The full access to the aviation market gave low-cost carriers (LCCs) the opportunity to establish an increasing number of crew and aircraft bases at a large number of secondary airports. Their strategy was opposite to the strategy of major airlines that remained at their hub bases. The accelerated growth of LCC's market share was aided by direct Internet booking platforms and by generating a completely new low-fare demand segment.

The average prices continued to decline in the period from 1993 to 2007 but started to rise again due to the rapidly rising cost of fuel (Goetz and Vowles 2009). In addition, airlines started to apply many tactics in pricing. Most notable was the introduction of ancillary fees[8] in addition to and separate from the price of airline tickets (Peterson 2018). This proved to be a very successful tactic due to the pronounced price sensitivity of large numbers of passengers who buy tickets based on the lowest price overlooking these extra charges.

However, this positive aspect was not equally applied everywhere. Many shorter-distance and lower-demand city-pair markets in the US experienced problems with airline service and prices. Due to the dominant position of regional airlines with market shares of 60% or higher, on the concerned markets passengers tend to pay higher prices and experience a decline in the quality of airline service, measured in increased congestion and delays, longer travel times, and poorer customer service (Goetz and Vowles 2009).

Competition between LCCs and major airlines resulted in reductions in labour costs, greater use of regional aircraft and network reconfiguration, and more flexibility in pricing adopted by full service airlines (Fu et al. 2010). However, along with these changes, the aircraft interior became less and less comfortable, and in-flight service was reduced for the average traveller. Consequently, the consumer market became highly segmented (Peterson 2018).

The last decade (2010–2020) indicates new trends in air transport. The low-cost market has shown some symptoms of saturation in certain markets or in some versions of the low-cost business model (De Wit and Zuidberg 2012, De Wit 2014). Some LCCs are shifting to more primary airports looking for new higher density, less seasonal, and higher yield routes, which may result in more direct competition between LCCs and full-service airlines (Burghouwt and De Wit 2015).

To facilitate the liberalisation and create a level playing ground in air transport markets, certain areas need to remain regulated or at least under the supervision of a relevant authority. To protect passengers and aircraft in the EU, national safety rules were replaced by common safety rules, which have gradually been extended to cover the entire aviation sector and to guarantee a uniform, high level of safety throughout the internal air transport market. At the international level, ICAO is responsible for setting minimum aviation safety standards, but these standards are not

[8] Additional fees are usually associated with the services such as checking-in a bag or choosing the seat.

obligatory and their application mainly depends on the member states' goodwill. For this reason, in the EU a European Aviation Safety Agency (EASA) has been established which, inter alia, is responsible for drafting regulatory provisions in this field. Moreover, security requirements at all EU airports have also been harmonised to better prevent malicious acts against aircraft and their passengers and crew. Furthermore, common rules (Regulation (EC) No 261/2004) to protect air passengers' rights aim at ensuring that passengers receive at least a minimum level of assistance in the event of serious delays or cancellation. These rules also provide for compensation schemes. An important aspect of the EU competition policy is the merger control aimed at further development of the internal market and prevention of those concentrations which may significantly impede effective competition in the common market (Council Merger Regulation (ECMR 04) No 139/2004).

Civil aviation has significant climate impacts. ICAO has been working actively towards developing a global solution to address CO_2 emissions from international aviation. In 2016, ICAO adopted the Resolution on the Carbon Offsetting and Reduction Scheme for International Aviation (CORSIA). The main goal of the Resolution is to compensate for the growth of international aviation emissions beyond 2020 levels through international offsets. In the next chapter, we will focus on the details of environmental regulation in aviation.

1.4.2 Summary

Deregulation of the air transport market was certainly successful in lowering average prices, increasing service quality in terms of frequency and other custom services, and increasing airlines' efficiency, while maintaining a good safety record. All these deregulation efforts resulted in significant traffic growth. In the EU alone, while the number of passenger-kilometres within the EU increased by around 23%, for air transport it jumped by about 74%, in the period from 1995 to 2014. Aviation's share in total passenger transport increased from 6.5% to 9.2%, which is by far the strongest growth of all modes of transport in the EU.[9] Moreover, airlines optimized their network structures and pricing strategies, which resulted in improved operational efficiency, higher load factors, and reduced average costs (Fu et al. 2010). In the period after the recovery from the 2008 economic crisis, the airline industry has become consistently profitable, globally and domestically, and this has led to the mass ordering of new airplanes from manufacturers (Peterson 2018).

[9] https://www.europarl.europa.eu/factsheets/en/sheet/131/luftverkehr-marktregelungen

Though deregulation allowed airlines to make their own decisions regarding networks, schedules, prices, mergers, and acquisitions, not everyone was equally successful at it. Many airlines made bad decisions resulting in large financial losses which led them into bankruptcy. During the regulation period, the airline industry did not face such financial losses anywhere like those of the post-deregulation period. Flight delays are increasing due to a lack of aviation infrastructure and the failure to allocate airspace capacity efficiently. These delays are very costly for airlines and passengers and amount to more than five billion dollars per year due to the increased costs for aircraft operation and ground personnel and loss of passengers' time (Button 2001).

The initial effect of deregulation towards removing barriers to competition was promising. But due to the oligopolistic nature of this industry, very soon many airlines began to merge, in a much higher concentration than in the pre-deregulation era (Pavlović and Babić 2018). Thus, there is a compelling need for at least some regulatory oversight over this industry, since it exhibits a tendency towards destructive competition (Goetz and Vowles 2009).

1.5 Global air transport development

Air transport is a very important component of the global transport system in terms of speed, quality of supplies, and geographic mobility. It provides the only worldwide transport network, which makes it essential for global business. Due to its global character, air transport has a great positive impact on global economic growth. This is especially reflected in the development of tourism, increasing the trade volume and increasing the mobility of people. Air transport is one of the transport sectors with a very high growth rate, and with a faster growth rate than the global gross domestic product (GDP) growth rate. According to the International Monetary Fund, in the period from 2018 to 2020, the trend of increasing world GDP was on average between 3.3 and 3.6%, on an annual level, while the increase in the air traffic volume was on average about 5%. Also, it is important to note that the world economy (e.g., economic crises, recession, and movement of air fuel prices) and world events (e.g., wars and other crisis situations) affect air transport development and operations. Thus, events such as the oil crisis of 1973, the war between Iraq and Iran in the 1980s, the Gulf War of 1990 and 1991, the financial crisis in Southeast Asia in 1997 and 1998, the terrorist attacks in the United States in 2001, the SARS epidemic in Asia in 2003 and the global financial crisis of 2007 and 2008, caused some very turbulent periods for air transport. Despite these downturns, world air traffic has shown stable long-term growth since the 1970s. The COVID-19 pandemic, which hit the whole world at the beginning of 2020, is the first crisis situation that caused a long-term disruption in the air transport market. The return to the 2019 level of air traffic and the continuation of the growing trend is uncertain. According to the current forecasts made at the beginning of 2021, the air traffic recovery cannot be expected before 2024 (IATA 2020a).

Until the beginning of 2020, the number of flights and the number of passengers in air transport increased year after year. The development of passenger air transport in the world, in the period from 2004 to 2020, is shown in Figures 1.4 and 1.5. In the period from 2010 to 2019, the average annual growth rate of the number of passengers was 6.2%. The total

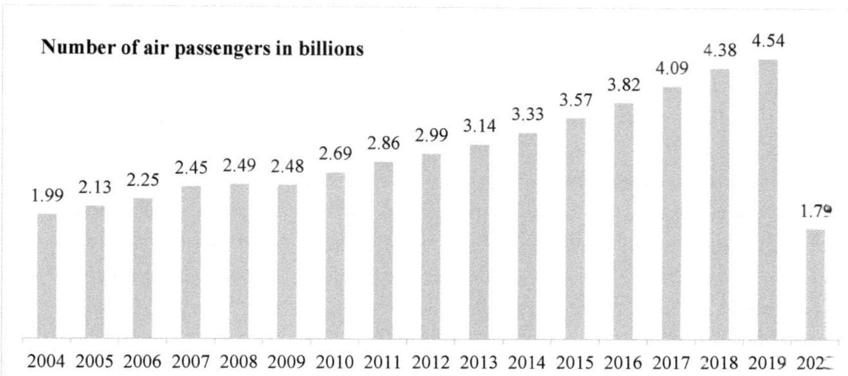

Figure 1.4. Annual number of passengers in the world (in billions), in commercial regular air transport.

Source: https://www.statista.com/statistics/564717/airline-industry-passenger-traffic-globally/

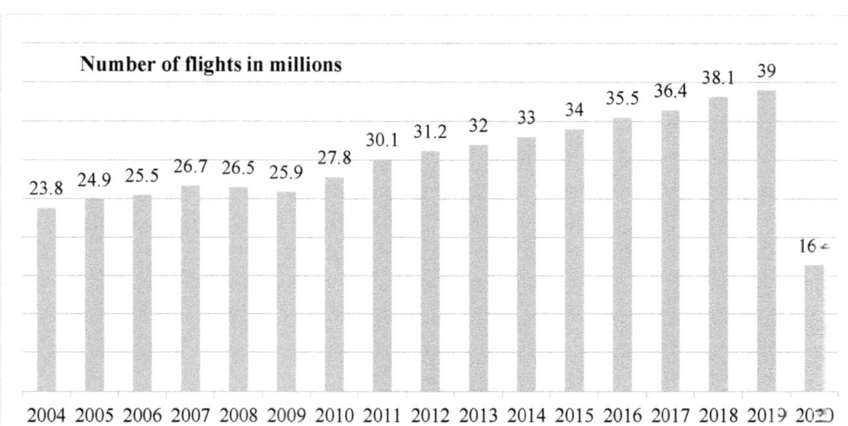

Figure 1.5. Annual number of flights in the world (in millions), in commercial scheduled air transport.

Data source: https://financesonline.com/number-of-flights-worldwide/

number of passengers in air transport in 2019 was 4.54 billion passengers (IATA 2020b). Due to the crisis caused by the COVID-19 pandemic, the number of air passengers in 2020 decreased drastically and, according to ICAO estimates, amounted to 1.79 billion passengers, which is 60.6% less than in 2019.

The number of flights increased from 23.8 million in 2004 to 39 million in 2019. This means that in 2019, more than 106 thousand commercial flights on average were performed every day and about 12.4 million air passengers were transported. In 2008 and 2009, there was a slight decline in the number of flights, which was caused by the global financial crisis in 2007 and 2008. In the period from 2010 to 2019, the number of flights grew by an average of 4.2% annually. In 2020, the number of flights dropped to 16.4 million, which is 57.9% less than the number of flights in 2019.

Long-term forecasts, which were made before the COVID-19 period, indicated a further increase in the volume of air traffic, up to 10 billion passengers in 2040. The increase in the number of passengers can be explained by the fact that the price of air transport is lower and this made it more available to the general population in comparison to the period 20 years ago. About 30% of air passengers in the world use the services of low-cost airlines, while in Europe the share of passengers who use low-cost airlines is even higher (36%). Moreover, rising global GDP, disposable income, and living standards also contribute to increasing the demand for travel for both business and leisure purposes. Given that air transport records exceptional growth within the transport sector and also in a broader framework, such as the regional economy, the question of balancing between positive and negative effects on society and the environment arises (Janić 2007).

The positive effects of air transport development are reflected in the global economy and people's daily lives, as well as in a large number of employees in the air transport sector. Air transport is a key element in the 'world's largest industry' of the last sixty years—travel and tourism (Kalić 2012). Directly in the field of air transport, about 11.3 million people are employed, of whom approximately 3.6 million work in airlines, more than 6 million at airports (airport jobs, customs, police, forwarding, restaurants, shops, etc.), 1.3 million people in the civil aviation sector in the production of aircraft systems, devices, engines, etc., and about 237 thousand people work in air traffic control agencies, etc. (ATAG 2020). Indirectly, air traffic also affects employees who produce or provide services within the energy sector and the information technology sector.

Negative impacts of air transport are related to environmental pollution, the noise made by airplanes, delays in air traffic that occur as a result of traffic congestion at airports and in the air. Although gas emissions are lower compared to other sectors, they are not expected to decrease in the coming years. Despite the fact that huge efforts are being made to design engines that are less noisy and consume less fuel, the level of pollution cannot be reduced because the number of flights is growing. The increase of the infrastructure capacity that can support this growth of traffic is also questionable in the future.

1.6 Green air transport system

Nowadays we are witnessing that the demand for aviation services on the global level has been increasing from year to year. The demand growth is stimulated by global population growth, as well as by increasing the welfare of societies and lowering the unit costs of air transport. With the increase of demand, the environmental influence of aviation increases, as well. Moreover, aviation and the environment are cross-interacting (Schmitt and Gollnick 2016): aviation produces emissions and noise affecting the environment, while weather conditions and natural disasters (rain, snow, ice, storm, thunderstorm, wind and turbulence, volcanic ash, etc.) influence aviation and could lead to operational restrictions. Humanity now faces the challenge to balance the need for aviation services and their negative impact. The negative impacts of aviation could be perceived through emissions, fuel burn, and noise impact, which pollute the environment. Each part of the air transport system can influence environmental pollution by performing usual activities, but all of the parts and activities are interacting with each other.

Airlines with their main resources—aircraft, produce noise that directly affects the quality of human life and emissions that further affect the atmosphere and influence climate changes. Moreover, aircraft indirectly influence land use. Precisely, aircraft performances, especially during the climbing and landing phases, define the required runway length at an airport, which is closely related to land use. Apart from that, local communities in the airport surroundings face the problem of noise and lower air quality, as well. Due to fact that airlines are able to order and buy/lease aircraft which are in use, a great responsibility is on the aircraft manufacturer. Therefore, in recent years, manufacturers have been putting an effort to make environmentally friendly aircraft, which also will be economically competitive on the aircraft market (Schmitt and Gollnick 2016). Land use in terms of different herbicides and pesticides used on airports to clear the airfield has certain impacts on the environment, as these can cause land pollution. Maintenance related activities such as de-icing or engine washing also contribute to pollution. Recycling of aircraft which cannot be used anymore (at the end of aircraft life) or any other technical system in air transport also becomes one of the relevant issues.

Airports have a wider spectrum of influences, but also more possibilities to support a sustainable system and to be environmentally friendly. Airports can provide energy savings by designing their buildings in away that can enable and support these savings. For example, if the orientation of the building and glass façade is suitable, ample quantities of daylight illumination could be adequately distributed into the internal spaces, even during cloudy days. Therefore, energy-efficient electric lighting is provided and energy can be saved in this way.

Due to environmental pollution and climate changes, a global tendency to prevent disasters is reflected in the fact that even green buildings are being built. One of the most widely used green building rating systems in the world is the Leadership in Energy and Environmental Design (LEED). LEED provides a framework for healthy, highly efficient, and cost-saving green buildings being globally recognized as a symbol of sustainability achievement and leadership. LEED is divided into subcategories depending on the project type: Building Design and Construction (BD+C); Interior Design and Construction (ID+C); Building Operations and Maintenance (O+M); Neighbourhood Development (ND); Homes; and Cities and Communities. For these categories, four levels of certification are available (LEED Certified, LEED Silver, LEED Gold, and LEED Platinum) depending on performances achieved in areas such as Location and Transport, Sustainable Sites, Water Efficiency, Energy and Atmosphere, Materials and Resources, Indoor Environmental Quality.

According to ICAO 2019, eco-design of airport buildings refers to considerations of green buildings, and environmental and resource-efficient operation and management of the airport building from a life-cycle perspective, including the overall process of airport planning, design, construction, operation, maintenance, refurbishment, and demolition. Many different issues related to the environment and sustainability need to be considered when planning airport buildings. Since airports are complex systems involving and hosting many stakeholders, it is expected that the sustainability of an airport combines economic, environmental, and social considerations into planning, design, construction, operations, and maintenance.

World's leading airports have been joining the initiative to support a greener environment and their buildings are awarded LEED certification. Some examples of LEED certified airports are: San Francisco International Airport (SFO) Terminal 2, Boston International Airport (BOS) Terminal A, Hartsfield-Jackson Atlanta International Airport (ATL) International Terminal, and Istanbul Airport (IST) as the world's largest LEED Gold certified building.

Moreover, coordination among different stakeholders in air transport (such as for example, Collaborative Decision Making—CDM, or introduction of multimodality in air passenger travels by the central position of air transport) can also contribute to reducing emissions. All coordination actions are based on the appropriate data sharing, which leads to improvement in the system performances, reduction of unnecessary delays and holdings, which further reduces noise and emissions, and can optimize the distribution of resources, as well.

In addition to green airports, manufacturers started to produce green aircraft that are interesting for airlines. Aircraft produce noise and emissions of harmful gasses whether in flight or on the ground. Regarding noise, it can be noticed that the main sources of aircraft noise are engine sources, which include fan, propeller, core, and jet noise, and airframe noise sources which encompass trailing edge, landing gear, slat, and flap (side-edge) noise (Nelson and Reddy 2017). However, there are also other sources of noise, such as air brakes or wing spoilers. To improve aircraft performances and to reduce noise levels, each of the noise sources can be improved and may enable noise reduction.

Also, the use of alternative fuels could reduce gas emissions, therefore, these fuels are being studied as replacements for current conventional-petroleum-derived aircraft engine fuels. One of the possible ways to gain multiple benefits is to improve aircraft fuel efficiency. Aircraft fuel efficiency could be achieved through engine design, and this will lead to a reduction of fuel consumption, decrease of environmental impact caused by emissions, and reduction of costs (Nelson and Reddy 2017). However, the key factors affecting the future of green aviation are technology trends, economic trends, policy and regulatory trends, and social trends (Nelson and Reddy 2017). Nowadays, we are facing a global aviation crisis caused by the global pandemic, as well as the challenge to forecast the future of aviation.

References

Air Transport Action Group (ATAG). 2020. Blueprint for a Green Recovery. https://www.atag.org/our-publications/latest-publications.html.

Boeing. 2021. Boeing History Chronology, https://www.boeing.com/resources/boeingdotcom/history/pdf/Boeing-Chronology.pdf Accessed: April, 2021.

Burghouwt, G. and de Wit, J.G. 2005. Temporal configurations of european airline networks. Journal of Air Transport Management 11(3): 185–198.

Burghouwt, G. and de Wit, J.G. 2015. In the wake of liberalisation: long-term developments in the EU air transport market. Transportation Policy 43: 104–113.

Burghouwt, G., Mendes de Leon, P. and de Wit, J. 2015. EU air transport liberalisation process, impacts and future considerations. Discussion Paper No. 2015-04. OECD.

Button, K. 2001. Deregulation and liberalization of european air transport markets. Innovation: The European Journal of Social Science Research 14(3): 255–275.

Cento, A. 2009. The Airline Industry: Challenges in The 21st Century, Springer.

Council Merger Regulation (EC) No. 139/2004 of 20 January 2004 on the Control of Concentration between Undertakings. 2004. OJ C L 24/1, 29.1.2004.

De Wit, J.G. and Zuidberg, J. 2012. The growth limits of the low cost carrier model. Journal of Air Transport Management 21: 17–23.

De Wit, J.G. 2014. An Unlevel Playing Field? Ah yes, you mean protectionism. Journal of Air Transport Management 41: 22–29.

Dempsey, P.S. 2008. The financial performance of the airline industry post-deregulation. Houston Law Review 45(2).

Fischer, W.J. 2018. Transportation economic regulation in practice. TR News, Number 315: 3–8.

Fu, X., Oum, T.H. and Zhang, A. 2010. Air transport liberalization and its impacts on airline competition and air passenger traffic. Transportation Journal 49(4): 24–41.

Goetz, A.R. and Vowles, T.M. 2009. The good, the bad, and the ugly: 30 years of US airline deregulation. Journal of Transport Geography 17(4): 251–263.

https://www.capgemini.com/wp-content/uploads/2017/07/Collaborative_Decision_Making_in_Aviation.pdf https://www.iata.org/contentassets/5c1a116a6120415f87f3dadfa38859d2/iata-acdm-recommendations-v1.pdf.

https://www.iata.org/en/programs/safety/.

https://www.icao.int/annual-report-2017/Pages/the-world-of-air-transport-in-2017.aspx.

https://www.icao.int/safety/Documents/ICAO_SR_2019_29082019.pdf.

https://skybrary.aero/articles/airport-collaborative-decision-making-cdm.

https://www.statista.com/statistics/564717/airline-industry-passenger-traffic-globally/.

Humpreys, I. and Francis, G. 2017. The airport-airline relationships. *In*: Budd, L. and Ison, S. (eds.). Air Transport Management: An International Perspective. Routeledge, Taylor & Francis Group, London and New York.

IATA. 2020a. Recovery Delayed as International Travel Remains Locked Down, https://www.iata.org/en/pressroom/pr/2020-07-28-02/ Accessed: May, 2021.

IATA. 2020b. Industry Losses to Top $84 Billion in 2020, https://www.iata.org/en/pressroom/pr/2020-06-09-01/ Accessed: May, 2021.

ICAO. 2018. Annex 1 Personnel Licensing, Twelfth Edition.

ICAO. 2019. The Eco Design of Airport Buildings, https://www.icao.int/environmental-protection/Documents/ecoairports/Final%20Airport%20Eco%20Design%20Overview.pdf Accessed: April, 2021.

Janić, M. 2000. Air Transport System Analysis and Modelling (Transportation Studies), 1st Edition, Gordon and Breach Science Publishers, Amsterdam, The Netherlands.

Janić, M. 2007. The Sustainability of Air Transportation: A Quantitative Analysis and Assessment, Taylor & Francis Ltd., Aldershot, England, United Kingdom.

Janić, M. 2013. Airport Analysis, Planning and Design: Demand, Capacity and Congestion, Novinka, Nova Science Publishers, Inc. New York.

Janić, M. 2021. System Analysis and Modelling in Air Transport: Demand, Capacity, Quality of Service, Economics, and Sustainability, CRS Press, Taylor & Francis Group.

Kalić, M. 2012. Planiranje Prevoženja I Eksploatacija Vazduhoplova1, Univerzitet u Beogradu - Saobraćajni fakultet.

Morrison, S. and Winston, C. 1986. The Economic Effects of Airline Deregulation. Washington DC: The Brookings Institution.

Nejhum, E. 2006. Letelice, Enciklopedija DK Sveznanje, KNJIGA Komerc, Politika NM.

Nelson, E.S. and Reddy, D.R. 2017. Green Aviation: Reduction of Environmental Impact Through Aircraft Technology and Alternative Fuels. 1st ed. CRC Press, Taylor & Francis Group, London, UK.

Nolan, M.S. 2011. Fundamentals of Air Traffic Control, 5th Edition, Clifton Park, N.Y.: Delmar Cengage Learning.

OECD. 2014. Air Service Agreement, Liberalisation and Airline Alliances. Inter VISTAS Consulting Inc.

Pavlović, D. and Babić, D. 2018. Recent trends in assessment of proposed consolidations in EU airline industry–From discretion to arbitrariness. Transport Policy 69: 65–77.
Peterson, R. 2018. Impacts of airline deregulation. TR News, Number 315: 10–17.
Petit, E. 1967. Histoire Mondiale de l'Aviation, Hachette.
Robyn, D. 2018. The unfinished business of transportation deregulation. TR News, Number 315: 49–54.
Schmitt, D. and Gollnick, V. 2016. Air Transport System, Springer Wien Heidelberg New York Dordrecht London.
Tonković, R. 2017. Nebeske Heroine Sveta, Vizartis, Beograd.
Transportation Research Board. 1999. Special Report 255: Entry and Competition in the US Airline Industry—Issues and Opportunities. National Research Council, Washington, DC.
US Air Transport Association (ATA). 2006. Annual Revenue and Earnings, US Airlines-all Services. ATA, Washington, DC.
US Department of Transportation. 2000. Changes in Pricing since Deregulation. Office of Aviation Analysis, Washington, DC.
Zorić, D. 1977. Osnovi Vazdušnog Saobraćaja I Transporta, Tehnički Školski Centar KoV JNA.

Chapter 2
Aircraft

2.1 Introduction

The aircraft represents the main technical resource of airlines. By using these means of transportation, airlines strive to perform safe, smooth, efficient, and effective passenger and cargo transport. Consequently, aircraft manufacturers tend to fulfil the needs and expectations of airlines and their passengers. Because the requirements are diverse, even some of them conflicting in the aircraft design process, designers and manufacturers handle trade-offs among them, while meeting all the necessary technical requirements. These requirements refer to safety issues, reliability[1] goals, and performance specifications. Reliability is an essential contributor to safety, and it is affected by design, development, production quality control, control of suppliers and subcontractors, and maintenance (O'Connor and Kleyner 2012). Aircraft design is a complex process that involves many different disciplines, such as aerodynamics, structural design, propulsion, on-board systems, flight mechanics, costs, and emissions assessment, to obtain a holistic approach and to derive a global "best" solution at the entire aircraft-level (Johansson et al. 2017, Fioriti et al. 2020). Designing an aircraft is a very long process (15 to 20 years) that encompasses different phases, from concept design, preliminary design and detail design, testing and refinement, to certification. Additionally, this process is extremely expensive. From the airlines' point of view, the most important issues related to a new aircraft are its performances, both economic and environmental. These performances are related to reducing

[1] Reliability is the ability of a component, process or a system to perform a required function correctly under stated condition within a given scope, during a given period (Isermann 2006).

fuel consumption, reduction of the mass of the structure, increased payload capacity,[2] and environmental issues.

Aircraft structures must be light but strong and stiff enough to withstand the various forces acting on the airplane at all stages of flight (take-off, climb, cruise, descent, and landing) and during its manoeuvres (turns). They must also be durable enough to withstand these forces during the entire period of use of the aircraft. Thus, all aircraft materials must be reliable, and have superior fatigue characteristics, and withstand a wide range of temperatures. Additionally, it is very important to use appropriate materials. In the case of commercial aircraft, the most common structural materials used are aluminium, steel, and titanium alloys and composites. As of recently, aluminium is increasingly being replaced by titanium and matrix/metal/ceramic/carbon composite materials. The selection of materials depends on the component shape, dimensional tolerance, mechanical properties, processing technology, cost, and availability of the material (Boyer and Padmapriya 2016). Despite the tendency to make planes as light as possible, the mass of the plane is not small. The mass of an empty aircraft used in commercial transport commonly varies from 10 tonnes (regional airplanes) to more than 100 tonnes (wide-body airplanes). The typical operating empty mass for the Airbus 380 (the largest passenger airplane) is 277 tonnes,[3] while its maximum take-off mass is 560 tonnes.[4]

The four forces acting on an aircraft in straight-and-level flight are thrust, drag, lift, and weight (Figure 2.1). Thrust is the force that moves an aircraft in the flight direction, provided by a piston, turboprop, or jet

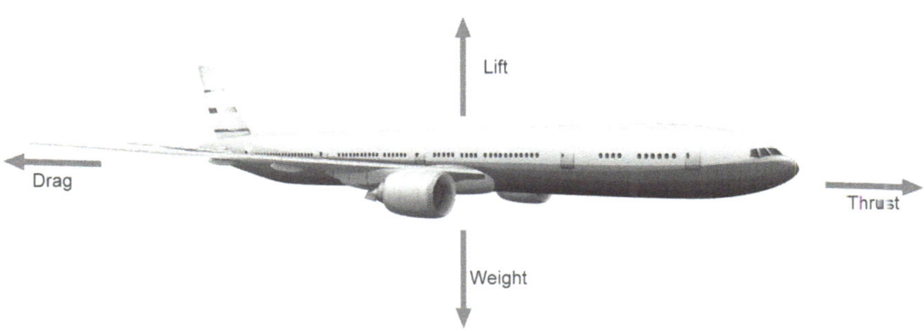

Figure 2.1. The four forces of flight.

2 The payload is related to passengers, baggage, cargo and mail, and represents the load that generates revenue. Total payload capacity available, in the passenger cabin (number of seats, space for hand baggage) and the cargo compartments (space for baggage, cargo and mail), for the carriage of revenue load taking into account all operational restrictions.

3 https://modernairliners.com/airbus-a380/airbus-a380-specs/

4 https://www.airbus.com/aircraft/support-services/airport-operations-and-technical-data/aircraft-characteristics.html

engine. Thrust itself is a force than that can best be described by Newton's second law. This forward force opposes the force of drag. Drag is the aerodynamic force component parallel to the direction of relative motion. This is a retarding force caused by the disturbance of airflow by the aircraft and its parts. In other words, drag tends to slow the motion of aircraft, and acts opposite to the direction of motion. Lift is a force that is produced by the dynamic effect of the air flow acting on the aerofoil. This force opposes the downward force of weight. Lift represents a component of the aerodynamic force, perpendicular to the direction of aircraft movement through the air, which is equal to or exceeds the weight. Weight is the force that pulls the aircraft downward owing to gravity. When the four forces of flight are balanced, a plane flies in a level direction. The plane climbs if the forces of lift and thrust are greater than gravity and drag. To descend, thrust must be reduced below the level of drag and lift below the level of weight.

According to Bernoulli's principle and the equation of continuity, the air flow above the top of the wing is moving faster and generates an area of low air pressure, while the air below the wing is moving slower and results in an area of high air pressure (FAA 2016). Thus, the pressure on the top of the wing is lower than the pressure on the bottom of the wing. The pressure difference ensures the creation of lift force. In that way, the wings enable the lifting of the aircraft and its flight through the air.[5]

The flight of the aircraft is performed at different altitudes in the atmosphere. From the moment of take-off to reaching the cruising flight level, the planes pass through the troposphere[6] (the lowest layer), and sometimes through the tropopause and the stratosphere (the second layer of the atmosphere). Cruising level is most common at altitudes from 8,500 meters (app. 28,000 feet) to about 11,500 meters (38,000 feet). The temperature of the troposphere decreases with altitude. The rate at which the temperature decreases is 6.5°C/km, therefore at an altitude of 10,000 meters the temperature is –56.5°C, Airbus 2002. The atmospheric pressure also decreases with altitude. The tropopause is an intermediate layer in which the transition to the conditions prevailing in the stratosphere occurs. In contrast to the troposphere, flight in the stratosphere is pleasant and without turbulence.

The chapter is organized as follows. Parts of an aircraft and its functions are explained in the first part of this chapter. Additionally, descriptions of the cockpit, passenger cabin, and cargo compartment are given. The chapter covers several aircraft issues that are essential to airline

---

5 https://www.faa.gov/regulations_policies/handbooks_manuals/aviation/phak/

6 In the troposphere, there are horizontal and vertical movements of air masses, thermal turbulence occurs, and also clouds, fog, precipitation (snow, rain, hail) are formed in this layer, as well as such phenomena as thunder and storms.

operations. The aircraft range will be considered, to regarding the payload capacity and amount of fuel required for a given flight. The mass of plane as well as aircraft balance is very important during flight (including during take-off and landing), because of its impact on the stability and performance of the aircraft. The importance of aircraft maintenance will be explained. The last part of this chapter provides the classification of aircraft.

2.2 Parts of an aircraft and their function

Aircraft consists of the following components (Figure 2.2):

1. wings,
2. fuselage,
3. empennage,
4. landing gear, and
5. powerplant.

Figure 2.2. Aircraft.

2.2.1 Wings

The wing is a very important part of the plane because it provides the aircraft flight. As the wing moves through the air, an aerodynamic force is generated, which consists of two components: lift force and drag force. Therefore, the basic task of the wing is to achieve the necessary lift force that enables the flight of the aircraft. In addition, the wing can accommodate appropriate equipment, fuel tanks, landing gear, and power plants. The fuel tanks are an integral part of the wing's structure or consist of flexible containers mounted inside of the wing (FAA 2016b).

The wing consists of different moveable parts which control the lift amount and air flow direction. These are aerodynamic devices or flight control surfaces that enable the pilot to control flight during take-off, climbing, descent, and landing. Additionally, the turns are performed by the pilot using these aerodynamic devices.

If the wing is cut with a plane that is parallel to the direction of the air flow and perpendiculars to the horizontal plane, an aerofoil is obtained. In other words, the cross section of the wing is the aerofoil. An aerofoil represents an aerodynamically shaped body that produces an aerodynamic reaction (lift) perpendicular to its direction of motion, for a small resistance (drag) force on that plane. An aerofoil is defined by the suction surface (camber upper surface), pressure surface (camber lower surface), the leading edge, the trailing edge, and the chord line, Figure 2.3. The chord is defined as the distance from the leading edge to the trailing edge and generally varies along the span (McCormick 1995). The leading edge is the first point of the aerofoil or the wing line that the air flow encounters. One part of the air flow passes over the upper and the other below the lower surface. The shape of the aerofoil takes advantage of the air's response to certain physical laws. Because of the aerofoil shape, the air flow rate is higher on the camber upper surface of the wing due to the longer the distance that it crosses. Therefore, an area of lower pressure relative to the camber's lower surface is produced on the camber's upper surface. The lift force is a result not only of pressure differences between upper and lower wing surfaces, but also of the angle of air stream attack, aerofoil shape, air density, and air stream velocity.

During certain stages of flight, especially take-off and landings, it is necessary to change the surface, shape, and inclination of the wing, which is achieved by moving certain parts of it. There are several types of control surfaces of the wing that belong to the primary or secondary aircraft flight control systems (FAA 2016c). These are ailerons, flaps, slats, spoilers, and air brakes (Figure 2.4).

Ailerons. The ailerons belong to the primary control system. The position of the aileron is on the trailing edge of the wing. Ailerons are used in pairs, one on each wing, and move in the opposite direction from each other, to control the aircraft roll. This means that the lift force increases on one wing and decreases on the other wing, which results in the aircraft rolling. This is the movement in one direction or another, around the aircraft's longitudinal axis.

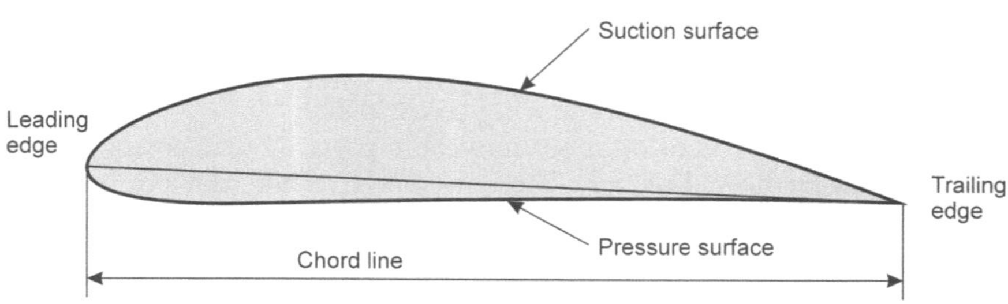

Figure 2.3. The aerofoil.

Flaps. Flaps are moving parts of the wing trailing edges located inboard (close to the fuselage), and belong to the secondary flight control system. Flaps are used to reduce the take-off and landing distances. When the airplane is taking off, the flaps help to produce more lift. Oppositely, flaps can provide a steep angle during landing. In this case, flaps are used to increase drag. Wing flaps change the shape of the airplane wing, by extending it (or retracting it) and providing the appropriate configuration for take-off, cruising, or landing.

Slats. Slats are parts on the leading edge of the wing. They move forward to increase the camber of the wing. Their purpose is to increase lift during low-speed operations such as take-off, initial climb, approach, and landing. Slats commonly have several possible positions, depending on needed configuration, and extend progressively simultaneously with flap extension, Figure 2.5 and Figure 2.6.

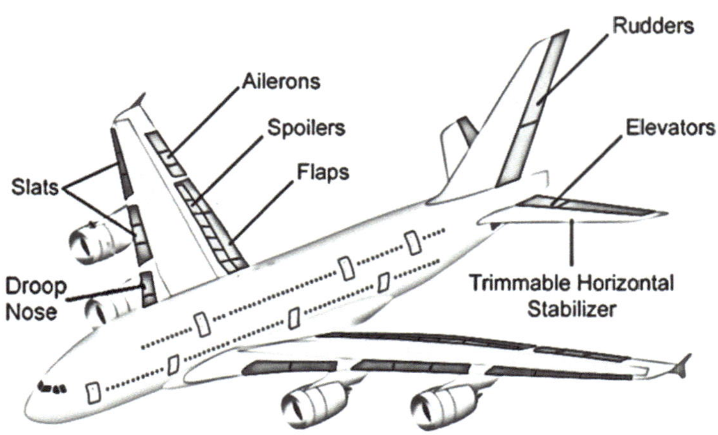

Figure 2.4. Parts of wing and tail.

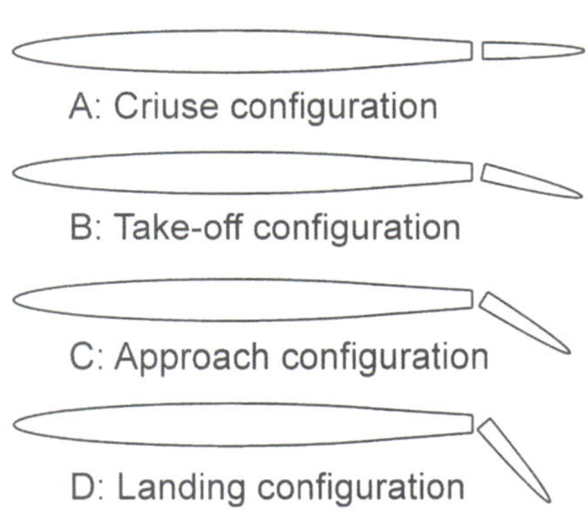

Figure 2.5. Positions of flaps during cruise, take-off, approach, and landing.

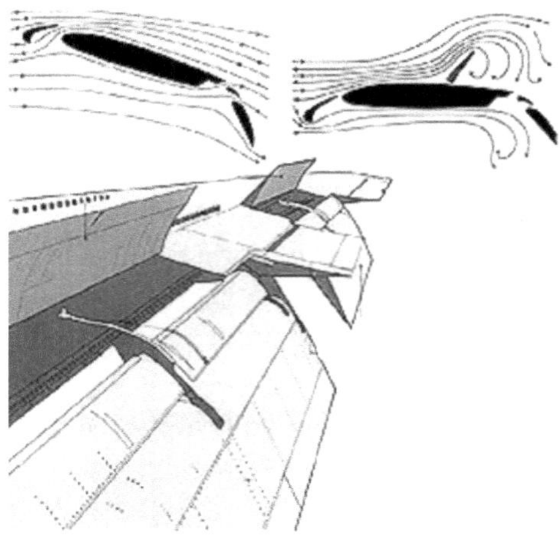

Figure 2.6. Wing obstruction by moving certain parts.
Source: https://www.modli.rs/flying/stw/stw4.html (10 December 2019)

Spoilers. In addition to flaps and slats, spoilers also comprise the secondary flight control system. Spoilers are metal plates hinged on the upper surface of the wings, and they act as airbrakes. The purpose of spoilers is to reduce lift and increase drag during descent and landing. Spoilers simultaneously increase drag and reduce lift. By reducing the lift-to-drag ratio after landing, spoilers transfer weight to the wheels, improving braking effectiveness. Additionally, spoilers enhance the effect of the ailerons, by disrupting the flow of air on the camber upper surface.

2.2.2 Fuselage

The fuselage is the main body of the aircraft, connecting the wings and tail assembly into the plane as a whole. It is tube shaped. The cockpit, passenger cabin, and cargo compartments are housed in the fuselage. From the aerodynamic point of view, the fuselage is a harmful element because it increases the force of resistance. That is why it should be as small as possible, but large enough for crew members to carry out their tasks without hindrance, and also for passengers to feel comfortable in the passenger cabin during the flight. On the other hand, from the commercial point of view, the fuselage is the most important part of the aircraft because it accommodates passengers, their baggage, and cargo (payload) providing airline revenue.

The fuselage as a shell structure is a construction consisting of transverse elements such as rings and frames and longitudinal elements—reinforced by stringers and longerons. An airplane has multiple openings such as passenger doors, emergency exit doors, and windows. The cross-section of the fuselage of airplanes with pressurized cabins is circular.

The cockpit and passenger cabin must provide appropriate environmental conditions. The aircraft cruising altitudes are between 8,000 and 12,000 meters. Thus, it is necessary that the cockpit and passenger cabin have conditions that are pleasant for pilots, cabin crew, and passengers. Human beings do not feel any difficulties and inconveniences to an altitude of about 2,500 m above sea level, while at higher altitudes the air becomes thinner and the effective percentage of oxygen decreases. Also, the atmospheric pressure and the temperature decrease with attitude. As previously mentioned, the rate at which the temperature decreases is 6.5°C/km, so at an altitude of 10,000 meters, the temperature is –56.5°C (–69.7°F). In such extreme environmental conditions, passengers would not be able to survive. Thus, it is necessary to provide a healthy and comfortable environment for passengers, cabin crew, and pilots and to pressurize the aircraft cabin, as well as to maintain thermal conditions. Therefore, cabin pressurization systems are designed to produce conditions equivalent to those at approximately 2,500 meters. In addition, the exchange of air from the inside of the aircraft to the outside is important, to remove carbon dioxide and water vapour, which are generated as a product of the breathing and perspiration of passengers.

2.2.3 Tail assembly

The tail assembly or empennage encompasses the horizontal tail and the vertical tail, Figure 2.7. Its purpose is to provide the stability and control of the airplane. The main task of horizontal tail surfaces is to balance the forces and moments of the wings in the dynamic system of forces of a moving aircraft. The control by the tail assembly is related to pitch and yaw.

The horizontal tail is composed of the horizontal stabilizer (or tailplane), which is a fixed part, and the elevator, which is a moveable part. The elevators, which are hinged to the rear of the horizontal stabilizer, allow the pilot to move the nose up and down. It provides control to

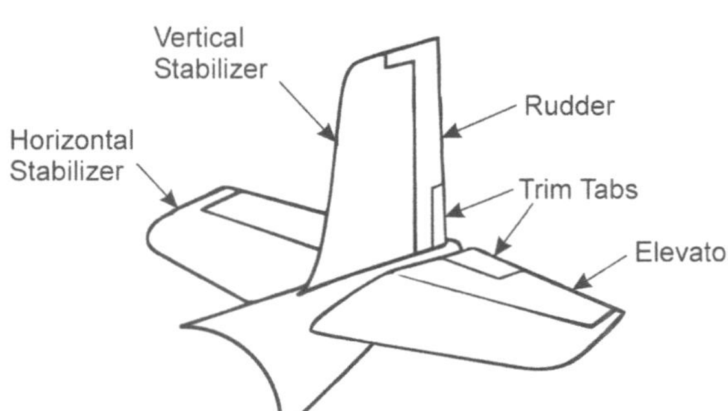

Figure 2.7. The tail assembly.

pitch the airplane about an axis parallel to the wing. The vertical tail is composed of the vertical stabilizer (or fin), which is a fixed part, and the rudder, which is a moveable part. The rudder, located on the rear edge of the fin, enables the pilot to control the airplanes' left-right movement (yaw control, around a vertical axis). There are also small moveable surfaces on the empennage—trim tabs. The trim tabs are located on the outer edges of the rudders and elevator and help the pilot to stabilize the airplane during flight, reducing the manual energy and workload of the pilot.

In terms of structural design, both horizontal and vertical tails are very similar to a wing. However, their structure is less complex than the structure of the wings, considering that they do not have as many moving parts.

2.2.4 Landing gear

The aircraft's landing gear is used for its movement on the ground from the moment of landing, until the next take-off. The undercarriage must provide longitudinal and transverse stability when the airplane taxiing on the ground. The landing gear of commercial airplanes is a tri-cycle undercarriage consisting of a single nose-gear forward and the two main gears under the wings area. This configuration is stable, tending to roll straight down the runway without yawing (McCormick 1995).

The landing gear has a complex structure because it must enable safe taxiing on the ground at high speeds (Figure 2.8). It has an extremely strong mechanism with systems that can withstand the ground impact of an aircraft weighing over 100 tons (the maximum landing mass of the

Figure 2.8. Landing gear. *Photo by Vladimir Trifunović.*

aircraft A380 is close to 400 t) at a speed of over 200 km/h. The landing gear consists of several components such as hydraulics systems (extraction and retraction mechanism), air/oil shock struts as an energy absorption component, brakes, wheels, and tyres. The landing gear must absorb the landing shock, which means absorption of the kinetic energy due to the aircraft mass and velocity of descent at impact. In order to keep the pistons and wheels aligned, most shock struts have torque links or torque arms hinged to the fixed upper and the lower cylinder (piston). This does not allow rotation of the shock absorber piston.

The landing gear has multiple wheel assemblies on each gear, to reduce the pressure on the runway and manoeuvring area of the airport, better braking, and easier retraction. In other words, multiple wheels spread the airplane weight over a larger area. The number and location of wheels on the main gear vary depending on the aircraft type and its mass. Further, the brakes are responsible for reducing the speed of the airplane and stopping it. In addition to high speeds, tyres must withstand heavy static and dynamic loads. During landing and braking, heat is also generated due to friction. Thus, aircraft tyres are one of the strongest pneumatic tyres ever produced. After take-off, the landing gear is retracted: the main gear into either the wings or fuselage and the nose gear into the forward fuselage, which reduces drag. It is only on the final approach that the pilot extends the landing gear.

2.2.5 Powerplant

The aircraft powerplant consists of an engine with all associated systems, which provides thrust to power the aircraft. The engine should be reliable, as small in size (dimensions and weight) as possible, efficient in terms of fuel consumption, and produce noise at an acceptable level. In addition, the engine should be as easy to maintain as possible. Large commercial airplanes must have at least two engines, because, in case of failure of one engine, the pilot can, following the appropriate procedures, land the plane safely. Generally, passenger aircraft have two, three, or four engines.

Three types of engines are used in large commercial transport and high-speed aircraft: turbojets, turboprops, and turbofans. Turboprop engines work in combination with a propeller to produce thrust, while turbojet and turbofan engines produce thrust by increasing the velocity of air flowing through the engine. These turbine engines differ in the path that the air takes through the engine and how power is produced. Turbofan engines work like a turbojet, but they have a very large ducted fan mounted on a shaft in front of the compressor (Taneja 1989). Turbofans were developed to combine some of the best features of the turbojet and the turboprop (FAA 2016d).

The most well-known manufacturers of turbojet and turbofan engines are General Electric Aviation (the CF6 for the Boeing 767, Boeing 747, Airbus A330, and GE90 (Boeing 777) and GEnx (Boeing 747-8, Boeing 787 Dreamliner)); Rolls-Royce, Pratt & Whitney, CFM International, etc. For turboprop engines, the leading manufacturer is Pratt & Whitney, namely engines PW127M (ATR 42, ATR 72), PW123C/D, PW123B/E, PW150 (De Havilland Canada Dash 8, series Q200, Q300, Q400, respectively).

2.3 Interior of the aircraft

As mentioned previously, the aircraft contains separate cabins for the flight crew and passengers, while the cargo is loaded into compartments located below the passenger cabin. This section describes the cockpit, passenger cabin, and cargo compartments.

2.3.1 Cockpit

Due to the cockpit being a vital interior section of the aircraft on the one hand, and part of the space that is non-commercial, the space should be used to the maximum, sufficient for the smooth operation of flight crew members and as small as possible. The flight crew (or aircrew) in civil public transport consists of at least two members (pilot–captain and co-pilot–first officer). The cockpit is designed to provide an appropriate arrangement of instruments and display positions and a good field of view for both the pilots inside the cockpit, as well as outside through the cockpit windows, Figure 2.9. To decrease the risk of negative outcomes, there are redundant systems for all crucial systems on airplanes, therefore, a large number of instruments are doubled, including two seats and two yokes or side sticks.

There are five groups of instruments in the cockpit: flight, engine, navigation, communication, and auxiliary instruments. Flight instruments provide the pilot with information on flight speed, acceleration, and

Figure 2.9. The cockpit. *Photo by Dieter Reisinger.*

aircraft position. This group of instruments consists of an air-speed indicator, an altimeter, a turn-and-bank indicator, a rate-of-climb indicator, which indicates the vertical speed, and an attitude gyro, which shows the position of the aircraft to the artificial horizon, etc. The engine instruments measure the performance of the engine and give indicators of various temperatures and pressures, engine speeds, and fuel flows.

The purpose of the navigation instruments is to follow the desired flight path and to determine the geographical position of the aircraft. The group of navigation instruments ranges from the simple magnetic compass and radiolocation to Global Positioning System (GPS) location devices, including standard navigation equipment such as Very-High Frequency Omnidirectional Range (VOR) and Non-directional Radio Beacon (NDB). The cockpit communication equipment consists of radio and audio panels with VHF (Very High Frequency), HF (High Frequency), SATCOM, which provides reliable, long-range voice and data communications, headsets, traffic collision avoidance system (TCAS), and transponders (which automatically send coded replies to interrogations by ATC radar). Auxiliary instruments include different instruments that control auxiliary systems such as electrical, hydraulic, and mechanical systems (e.g., the position of landing gear or cabin pressure and temperature) (Taneja 1989).

Special attention is paid to pilot comfort, the instrument panel arrangement, and the position of other necessary equipment. The cockpit must have adequate lighting, ventilation, and oxygen supply equipment. Also, the cockpit should have an emergency exit in case of an accident. The cockpit is separated from the passenger cabin, and cockpit doors were additionally secured after the terrorist attack that took place in the United States on 11 September 2001.

2.3.2 Passenger cabin

Behind the cockpit is the passenger cabin (Figure 2.10). The passenger compartment can be divided into several classes, for example, first, business and economy class. The difference between the classes is reflected in the characteristics of the seats (width, comfort, and the possibility of converting the seat into a bed), the space between the seats (longitudinally and laterally), and the level of services provided to passengers during the flight. The higher class means more space, more comfortable seats, and a higher level of service during the flight. The cabins of the different travel classes are often divided by curtains.

The first class section is located in the front of the passenger cabin and usually has a small number of seats. This is class with the best service, including personal in-flight entertainment systems, high quality food and beverages, and personalized service. The seats are extremely comfortable and can convert into beds. The space for each passenger is designed to

Figure 2.10. The passenger cabin. *Photo by Milica Kalić.*

provide privacy. The business class section is located between the first class section and the economy section. This class has more space, comfort, better service, and offers more amenities to passengers than the classes below it. However, compared to the first class, the service and comfort are at a lower level. The economy class primarily features the lowest level of service. The short longitudinal distance between seats, a large number of seats that are simple and of smaller width, and a smaller variety of food, beverages, and entertainment are features of the economy class.

Aircraft with only one class (economy) or aircraft with two classes (business and economy) are often used in short-haul and medium-haul traffic, depending on the type of airline. Most often, the largest number of seats is in economy class. For example, the Austrian Airlines fleet includes the Airbus 320-200 aircraft type. This aircraft has 30 rows of seats, with the first seven rows in business class and the rest in economy class. In business class, there are four seats in a row, two seats on each side of the aisle, and therefore there are a total of 28 seats in business class. In economy class, there are six seats in a row, three on each side of the aisle, which means that there are a total of 138 seats. In long-haul traffic, larger capacity aircraft are used, which can have up to nine seats per row (for example, a Boeing 777 has two seats by the window, five in the middle,

Figure 2.11. Hand luggage store place.
Source: Airbus, https://www.airbus.com/aircraft/passenger-aircraft/cabin-comfort/airspace/a320neo-airspace.html

and two by the opposite window, separated by aisles) or ten seats in a row (for example Boeing 747 has three seats by the window, four in the middle, and three by the opposite window).

The passenger cabin must be equipped with a space for storing hand luggage, which is located above the seats, along the entire length of the passenger cabin, on both sides of the aisle, Figure 2.11. For safety reasons, i.e., for the safety of the passengers, these spaces must be closed before take-off. If these spaces are full, and there is still hand luggage, then smaller hand luggage must be placed under the seat in front of the passenger. All seats must be equipped with oxygen masks, in case of loss of pressure in the cabin, and life vests, in case of emergency landing on water. At the back of each seat, there is a folding table for the passengers to use, as well as a pocket containing a manual with information related to the plane and emergency procedures, the airline magazine, brochure with prices for food, beverages, duty free shop products, etc.

In addition, there are spaces for cabin staff (kitchens and buffets) and toilets for passengers. There must be at least two doors for the entry and exit of passengers, and auxiliary exits for accidents or incidents. There must also be an exit on the roof of the cabin in case of landing on the water. According to ICAO, the interior layout of the cabin and the position and number of emergency exits, including the means of locating and illuminating the escape paths and exits, has to help and provide rapid evacuation of the aircraft in conditions likely to occur following an emergency landing.[7] Thus, the aircraft must be equipped with evacuation slides, which are used in case of emergency. There are different types

---

[7] https://www.icao.int/safety/fsix/Library/Reference_ANNEXES_for_ICAO_Ramp_Inspections_Guidance_2009_07_02.pdf

of slides (inflatable slide, inflatable slide/raft, inflatable exit ramp, and inflatable exit ramp/slide), which provide a nominal exit rate of 70 people per minute per lane. Inflatable life rafts must be part of the emergency and survival equipment in case of landing on water.

The cabin crew consists of a purser and flight attendants. The purser is the cabin crew chief on the flight. Generally, the cabin crew provides "routine services and respond to emergencies to ensure the safety and comfort of airline passengers".[8] The main tasks of the cabin crew are to greet passengers at the entrance to the aircraft, to assist passengers if necessary to find a seat on the aircraft and to store hand luggage. When boarding is completed, they check that the number of passengers corresponds to the number reported by airport services, close the door of the plane and inform the captain that all passengers have boarded. Before take-off, the cabin crew is required to provide passengers with basic safety-related information. After that, it checks whether all passengers have fastened their seat belts. During the flight, they serve food and beverages. In the case of turbulence during the flight, flight attendants must ensure the cabin is secure. Before landing, flight attendants instruct passengers on landing procedure, check that the seatbacks are in the upright position and that seat belts are fastened. A final cabin check must then be completed before landing. Upon landing, flight attendants must stand at exits and monitor the airplane and cabin as passengers disembark the plane.

The cabin crew is trained for emergencies and rescue, emergency evacuations after an emergency landing, and also for providing first aid to passengers. In addition, they are trained for situations such as smoke in the cabin, fires, depressurization, on-board births and deaths, dangerous goods and spills in the cabin, hijackings, and water landings. The number of cabin crew members depends on the type of aircraft in terms of capacity (number of passengers), number of classes on the aircraft, and duration of the flight.

2.3.3 Cargo compartments

The cargo compartments are located beneath the passenger cabin, in the lower half of the airplane. Large aircraft have more luggage compartments, both in the front and in the rear of the plane. As can be seen from Figure 2.12, cargo containers and ULDs,[9] as well as passenger luggage, can be placed in the compartments. Pressurisation is also provided in these spaces, so live animals can also be transported, in accordance with

[8] https://www.bls.gov/ooh/transportation-and-material-moving/

[9] Unit Load Devices (ULD) are used as containers for baggage and cargo carried in the holds of suitably dimensioned and equipped aircraft, and are secured so that they cannot move within the hold in flight (SKYbrary, https://www.skybrary.aero/index.php/Unit_Load_Devices_(ULD)).

Figure 2.12. Loading cargo and baggage on an aircraft.
Source: Flughafen München, https://www.munich-airport.com/electric-vehicles-at-the-airport-5938664

the airline's regulations. There is an exception for small animals. Pets, small dogs, and cats, up to 8 kg can be transported in the passenger cabin, housed in an appropriate container.[10]

2.4 Aircraft masses

A very important topic concerning airplanes is aircraft masses. When loading goods and embarking passengers, crucial attention should be paid to the total mass of the aircraft to prevent the violation of the airplane mass limit that the aircraft structure would not be able to withstand. The maximum mass that an aircraft may have at take-off is greater than the maximum mass that the aircraft may have at landing. The lower landing mass is achieved by fuel consumption. Also, the range is a function of the mass of the aircraft, the mass of payload, and the mass of fuel required for the realization of the flight. It is clear that for a longer range, the aircraft requires more fuel. For each type of aircraft, it is defined what is the maximum range (the maximum distance that the aircraft can cover from take-off to landing) with the maximum payload, and also what is the maximum range if the airplane tanks are full, where there is a limit on the mass of the payload that can be on the plane. To explain the payload range diagram, it is necessary to understand the definitions of different masses of aircraft.

[10] The total mass of pet and container (with ventilation net) may not exceed 8 kg (https://www.lufthansa.com/be/en/travelling-with-animals).

The manufacturer provides the aircraft operator with the empty mass of the aircraft. The manufactures' empty mass (MEM) represents the sum of the masses of the structure, power plant, furnishings, installations, systems, and other equipment that are considered an integral part of an aircraft without additional operator items. The Operating Empty Mass (OEM) is the mass of the aircraft, including the mass of the empty aircraft (aircraft structure, power plant, equipment, systems), the mass of standard items (electronic equipment, basic emergency equipment, structure buffet carrying containers for beverages and food, unusable fuel, oil and fluid at the bottom of the tanks, water for drinking and washing that cannot be spent, water in toilets and appropriate chemicals that are added to this water), the mass of operating units including operator items and the flight and cabin crew with their baggage, and other items needed to be taken on the flight.

The payload (P/L) is the mass of passengers and their baggage, cargo, and mail. Trip fuel consists of fuel for take-off, climbing, cruising, descent, and landing. Taxi fuel must be added to the trip fuel. A fuel reserve must be added for each flight, in addition to trip and taxi fuel. The amount of reserve fuel is not a function of the distance the plane needs to fly. According to regulations, for safety reasons, the aircraft must always load a certain amount of reserve fuel. The reserve fuel can be used in different situations, such as the flight in holding due to congestion at the destination airport, for the flight to the alternate airport, because of very bad weather or other disruption at the destination airport, pressurization failure, and/or engine failure during the flight. It is also possible that there will be higher fuel consumption due to navigation error or a change to the planned route, due to bad weather. Flight planning should take the following conditions into account for every flight: realistic airplane fuel consumption data, anticipated mass, expected weather conditions and air traffic services, and procedures and restrictions (Airbus 2002).

When the holding is required by air traffic controllers, it is generally flown on a "race track pattern", composed of two straight legs plus two 180 degree turns (Airbus 2002). The pilot received information from ATC about the holding at a certain altitude, at which the airplane will fly in a circle and wait until it is their turn to land.

The set of alternate airports is defined for each airport. Thus, the alternate fuel, as a part of reserve fuel, is the amount necessary to fly to the most distant alternate airport. In some cases, an alternate airport cannot service certain types of aircraft, due to insufficient runway length or lack of appropriate equipment. Accordingly, depending on the type of aircraft and the ability of the alternative airport to serve it, a flight will be rerouted to an appropriate alternate airport.

The Maximum Structural Zero Fuel Mass (Maximum Zero Wing Fuel Mass—MZFM) is the sum of the Operating Empty Mass (OEM) of an aircraft and the mass of the maximum payload (max P/L). The Maximum Design Landing Mass (MLM) is the maximum mass that an aircraft may have when landing, due to design limitations. The Maximum Design Take-off Mass (MT/OM) is the maximum mass an airplane may have at take-off (Figure 2.13). As already mentioned, the maximum design landing mass is lower than the maximum design take-off mass:

$$MLM < MT/OM$$

All the above-mentioned masses are provided by the aircraft manufacturer, and they represent the structural mass limits based on aircraft maximum structural capability established during aircraft design and certification (Table 2.1).

The structural masses are determined to achieve the performance provided in the aircraft manual, for the given structure and power plant. There are also operational masses such as operational take-off mass and operational landing mass (OT/OM and OLM). It is a mass that is less than the structural one, and it is limited by the length of the runway, the altitude of the airport, obstacles in the direction of take-off, and the local conditions at an airport. The mass that an aircraft has on take-off or landing is the actual mass. There is also an actual zero fuel mass. Each of these masses is less than or equal to the operational mass, as a result of the available payload on a particular flight and flight duration.

$$MT/OM \geq OT/OM \geq AT/OM$$

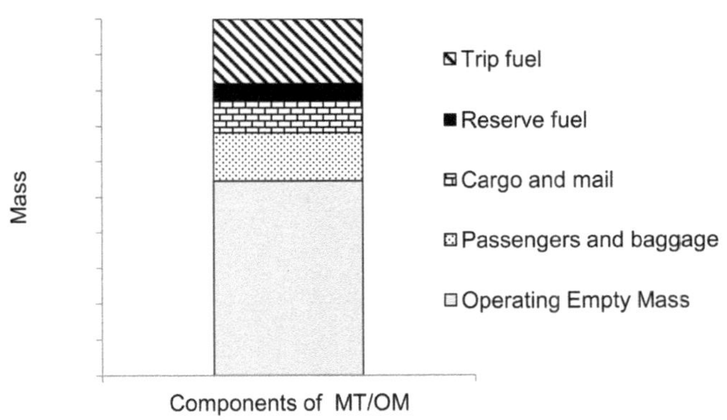

Figure 2.13. Take-off mass.

Table 2.1. Structural, operational and actual mass of an aircraft.

Mass	Structural mass	Operational mass	Actual mass
Take-off mass	MT/OM	OT/OM	AT/OM
Landing mass	MLM	OLM	ALM
Zero wing fuel mass	MZFM	-	AZFM

The structural take-off mass has the highest (maximum) value, the take-off operational mass can be equal to the structural mass but usually is less, while the actual take-off mass can be equal to the operating mass, but is usually less.

Before embarking passengers, loading luggage and cargo, and refuelling, the mass of the empty aircraft is the operating empty mass. The actual zero fuel mass (AZFM) represents the mass of embarking passengers and loaded cargo, baggage, and mail, which is added to the OEM. The AT/OM is obtained as a sum of AZFM and the amount of fuel needed for a given flight, Figure 2.14. The fuel consists of the taxi, trip, and reserve fuel.

During taxi-out to the runway, the airplane consumes a certain amount of fuel (taxi fuel). From take-off at the departure airport to the landing touchdown at the destination airport, the total amount of fuel in the tanks is reduced, because trip fuel is used, and possibly even part of the reserve fuel, for example for the holding. At the moment of landing, the aircraft has the actual landing mass (ALM), Figure 2.15.

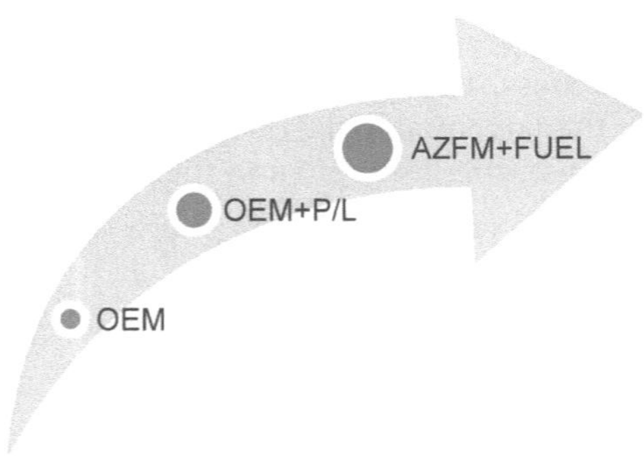

Figure 2.14. The mass that the aircraft at the moment the engine is started.

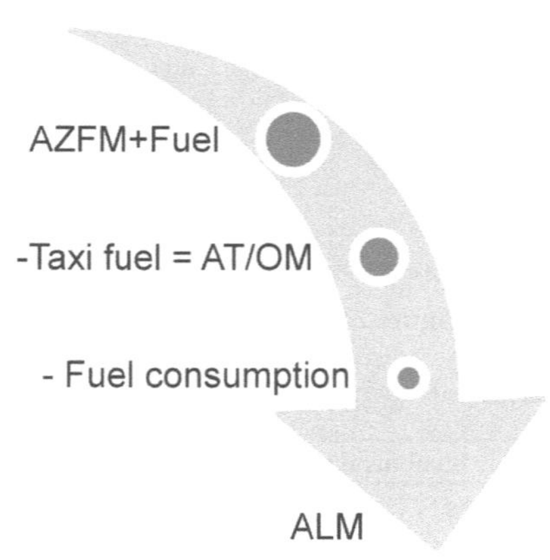

Figure 2.15. The mass that the aircraft has from brake release at the departure airport to the landing touchdown at the destination airport.

The example: **Airbus A319**
(*Source*: http://mediawiki.ivao.aero/index.php?title=Standard_aircraft_weights)

Manufacturer's Empty Mass (MEM)	36,779 kg
+ Operator's Items	+ 5,205 kg
= Operational Empty Mass (OEM)	= 41,981 kg
+ Payload (P/L)	+ 13,529 kg
= Actual Zero Fuel Mass (AZFM)	= 55,510 kg
+ Fuel	+ 13,239 kg
	= 68,749 kg
– Taxi fuel	– 100 kg
= Actual Take-Off Mass (ATOM)	= 68,649 kg
– Fuel consumption	– 10,900 kg
= Actual Landing Mass (ALM)	**= 57,749 kg**

2.5 Payload-range diagram

The payload-range diagram shows the relationship between the payload (masses of passengers, baggage, cargo and mail) and range for an aircraft, Figure 2.16. The x-axis represents the range or distance that the airplane can fly. The y-axis shows the masses. In Figure 2.16, the following masses are shown on the y-axis: OEM, MZFM, MLM, and MT/OM. The origin in Cartesian coordinate system represents zero distance and operating

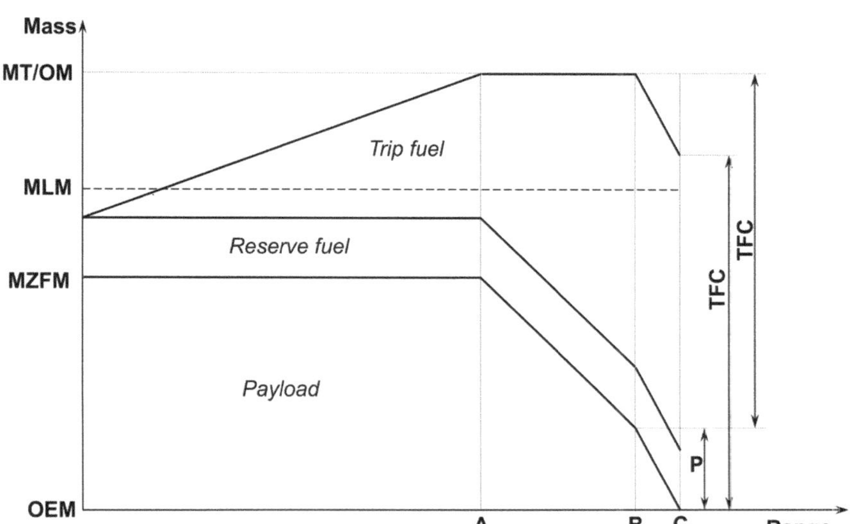

Figure 2.16. Payload range diagram.

empty mass of the aircraft (0, OEM). This diagram shows the effects of the trade-off between payload and fuel on the range of the aircraft. When planning a flight, it is necessary to know the limits related to payload (passengers, baggage, cargo and mail) that can be loaded without the plane being overloaded. On the other hand, it is clear that if the plane needs to fly a greater distance, it should also to plan for more fuel. As we can see from Figure 2.16, the maximum payload is limited by the maximum zero fuel mass (MZFM). When considering ranges from 0 to A (see x-axis), it is apparent that by increasing the amount of fuel a higher range can be achieved.

Range A is the largest distance to which the maximum payload can be transported, in compliance with the regulations that there is also reserve fuel on board, in addition to trip and taxi fuel. The diagram shows that the amount of reserve fuel is constant, to the distance that the aircraft needs to fly. Therefore, at range A, the aircraft can carry the maximum payload, in the amount of max P/L = MZFM - OEM, and fuel in the amount of MT/OM - MZFM. The label TFC (Total Fuel Capacity) indicates the capacity of the fuel tank. Based on the diagram, we can conclude that at range A, the fuel tank is partially filled.

If more fuel is carried, an even greater range can be achieved, but the problem is the maximum take-off mass. The diagram shows that between flights ranges A and B, the aircraft reaches the maximum take-off mass MT/OM, but the payload is less than the maximum payload. Accordingly, for ranges between A and B, an increase in the distance that the aircraft can be flown is made possible by reducing the amount of payload that can be transported.

The maximum distance that the aircraft can fly, while respecting the regulations on carrying reserve fuel, is possible when the fuel tanks are full. For range B, the total payload cargo that can be transported is P (see Figure 2.16). Therefore, **range B** is the maximum range that can be achieved with an aircraft when its total tank capacity is filled while respecting the regulation on carrying reserve fuel. For both range A and range B, the aircraft has a maximum take-off mass (MT/OM), with the fuel tanks being full for range B, and the payload being significantly lower than for range A.

In the case of the aircraft performing a ferry flight (flight without passengers or cargo), then a longer range can be achieved than the range B, and upon arrival at the given airport (**range C**) the plane will have the reserve fuel required by regulations. Figure 2.16 shows that the trip and taxi fuel for range B and range C is the same, explaining that during the whole flight without payload, the aircraft has a lower mass, and therefore the specific fuel consumption is lower, which increases the range.

Theoretically, if the total reserve fuel was consumed, the aircraft would achieve an even higher range—range E, which would be an emergency situation. This case is against the regulations and cannot be indicated in advance of the flight plan.

Payload is limited by mass and space. As already mentioned in the previous text, up to range A payload is limited by mass - MZFM, i.e., max P/L = MZFM - OEM. For flights beyond range A, the payload is limited by the mass MT/OM, so the allowed mass of payload that can be carried by aircraft is equal to P/L = MT/OM - FUEL - OEM. For example, the max P/L for A319neo is 17.7 tons, MT/OM is 75.5 t, while MLM is 63.9 t The capacity of the fuel tank is up to 26.7 t.

The payload limited by space is related to the volume of the baggage, cargo, and mail that can be accommodated in the cargo compartment and the number of passengers that can be accommodated in the passenger cabin. The number of passengers in the cabin must not exceed the number of seats on the aircraft.

It can happen that the plane is full (all seats are occupied) and that the cargo compartments are full (filled with light bulky goods), and that the total mass of payload is much less than the max P/L. In this case, although the mass limit has not been reached, no additional payload can be loaded into the aircraft, because the volume of the compartment has been used as well as the seats in the passenger cabin have been occupied (Razumenić 1984). For the A319neo, the maximum number of seats is 160. In the case of a two-class configuration, this type of aircraft can have between 120 and 150 seats. The total volume in the cargo compartment is 27 m^3.

2.6 Centre of gravity of the aircraft

Each part of the aircraft has its weight. The weights of all parts of the aircraft represent a system of bound parallel forces, and their resultant is the weight of the aircraft concentrated in one point. This point is called the centre of gravity of the aircraft (CG). In other words, the centre of gravity of the aircraft represents the imaginary point through which the force of gravity acts (G = mg). The position of the centre of gravity of the aircraft is of vital importance for all phases of the aircraft's flight, from take-off, climbing, cruising, descent to landing.

Depending on where the centre of gravity is located, the aircraft will have certain flight characteristics, such as its longitudinal stability and controllability. The position of the centre of gravity shifts during the flight for several reasons, due to (Razumenić 1984):

1. fuel consumption,
2. retraction and extension of the landing gear,

3. the movement of cabin crew members and passengers during the flight,
4. serving food and beverages during the flight, due to moving the trolley down the aisle, and
5. sale of the duty-free products to passengers during the summer, also due to the movement of the trolley down the aisle.

For this reason, aircraft manufacturers provide customers with limit values on the position of the centre of gravity within which flight is safe. This interval of recommended values is defined by the most forward and the most rearward permissible position for the centre of gravity. These limit values are given depending on the weight of the aircraft.

It should be noted that since the airplane is symmetrical, the position of the centre of gravity along the y-axis for the airplane is irrelevant (Figures 2.17 and 2.18). Minor deviations are possible, but this does not affect the flight characteristics of the aircraft. The position along the vertical axis (z-axis) is also irrelevant for the aircraft. Therefore, the position of the centre of gravity is expressed along the longitudinal axis (x-axis).

The centre of gravity is expressed in relation to the mean aerodynamic chord, abbreviated MAC, which is equal to the chord of an imaginary rectangular wing that has the same surface as the wing under consideration and which are affected by the same moments due to aerodynamic forces.

The position of the centre of gravity of the aircraft is expressed as a percentage of the mean aerodynamic chord, according to the relation:

$$CG = \frac{x - d}{MAC} * 100\%$$

where d is the distance of the leading edge of the MAC from the origin of Cartesian coordinate system, x is the position of CG on x-axis and MAC is length of the mean aerodynamic chord.

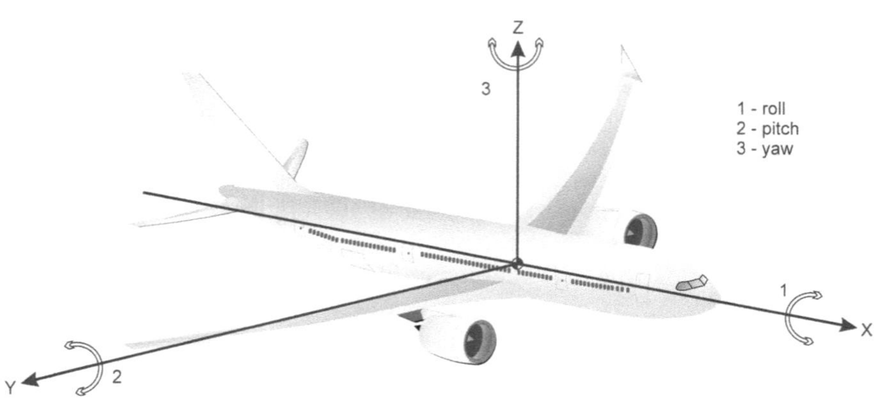

Figure 2.17. Aircraft in Cartesian coordinate system.

Figure 2.18. The position of CG and the mean aerodynamic chord in the horizontal plane.

As mentioned before, the allowable location of the CG is determined as the range defined by the most forward and the most rearward position. When the CG is out of the interval and its operational limits, the safety consequences can be extremely critical. If the centre of gravity were further forward, the aircraft would be very longitudinally stable and it would be very difficult to control it. Aircraft control refers to the ability of a pilot to change the flight conditions. If the centre of gravity were further back, the aircraft would be longitudinally unstable and it would be very easy to control. If the centre of gravity is even further back, the aircraft would tend to dive or climb progressively into a very steep dive or climb, or even a stall. In such a situation the pilot would not be able to control the aircraft.

There are certified and operational limits to the CG positions. Certified CG position limits are the values for which the aircraft has been tested to meet safety requirements in flight. Operational limits are provided by the aircraft manufacturer. The operational limits of the CG position cover a narrower range compared to the certified limits, thus increasing the level of safety. Unfortunately, errors can occur when entering data in the official document called the load sheet, or when loaded cargo is improperly arranged, or if the cargo is not secured in the cargo compartment, etc. These errors can produce unstable conditions during flight, causing a safety threat.

In terms of stability and controllability, safety is provided if the centre of gravity is within the operational limits. However, there are also recommendations for the position of the centre of gravity, when the aircraft has the best performance during the take-off and landing phases. For example, for the A320 aircraft, the recommendation is 25% MAC, Bonnet 2009. The load sheet contains all the necessary information related to the loaded cargo and passengers on board the aircraft. This list is an official document based on data related to number of passengers (per category) and loaded baggage, cargo and mail (mass per compartment). There are different methods for determining the loaded masses of payload and CG of an aircraft, such as the computational method, methods based on graphs and tables provided by the aircraft manufacturer. Based on data from the Loadsheet, but also the data on the total amount of fuel needed for the flight (taxi, trip, and reserve fuel), the position of the centre of gravity is calculated, as is the interval in which it will move during the flight. Sometimes it is necessary to enter changes or additional input in the load sheet after completion. These are related to last minute changes. Before take-off, the mass and balance documentation must represent the actual loaded state of the aircraft. There is a large amount of data in the load sheet, and additionally there are responsible persons who participate in the entire process, so mistakes are possible, which unfortunately occurs in practice. The reasons are numerous, and some of them are short time available for performing tasks or insufficiently trained staff.

2.6.1 Weighing aircraft

When determining the position of the centre of gravity of the aircraft, it is necessary to have accurate information about the operating empty mass of the aircraft. Therefore, the aircraft must be weighing from time to time, in accordance with the regulations of the civil aviation authorities. This can be, for example, every three years, or after major work on the aircraft.

The aircraft must be weighed because of changes that can be classified into two groups (Razumenić 1984):

1. changes related to the use of aircraft and which mainly result in an increase in mass (due to the application of paint when painting the exterior or interior of the aircraft, the collection of dust and moisture in inaccessible parts, etc.).
2. changes related to various modifications on the aircraft (new equipment, installation of additional seats, etc.).

Some modifications can increase the aircraft mass, while some can decrease it. The aircraft must be weighed indoors, to eliminate the influence of wind. Therefore, it is done in hangars, without air fluctuation and at a constant temperature.

2.6.2 Aircraft loading and balancing

Before each flight it is necessary to load cargo and baggage correctly (in terms of spatial distribution) and secure them, to prevent their movement within compartment during flight. This is done by the airport/airline ground handling service or outsourced operator. Since flight safety directly depends on this process, these tasks must be performed by highly responsible and trained staff. The data obtained by weighing the aircraft represent the starting point in the procedure of aircraft loading and balancing. The most important task is to properly load the cargo and to properly distribute the passengers in the passenger cabin (if the number of passengers in the cabin is significantly lesser than the number of seats on board). In addition to balancing, it is necessary to check whether the maximum, structural or operational mass of the aircraft has been exceeded by loading the aircraft at the airport.

During this procedure, many elements must be taken into account: aircraft registration number, aircraft version in a given situation, number of crew members in the cabin and passenger cabin, type of service provided for the given flight, special equipment on board, the amount of fuel and reserve, how many passengers are on board (by categories: adults, children, babies), the amount of cargo, mail, and baggage, as well as hand luggage in the cabin and what changes occurred at the last moment.

The standard weights of passengers, classified by categories (adults, children, and babies), are taken into account in the calculation of the centre of gravity position. Since the average mass of passengers is increasing, research is being done in this field as well, where, in addition to the category of passengers, gender is also taken into account. EASA (2009), according to the Survey on standard weights of passengers and baggage, recommended the following standard masses, 88 kg for adults (male passengers 94 kg, female passengers 75 kg), as well as standard mass for checked baggage 17 kg.[11]

2.7 Aircraft maintenance

To keep aircraft safe and operational, an airline needs to provide appropriate maintenance. Airlines may have their own maintenance department, but today the outsourcing of a non-core activity such as maintenance, ground handling, ticketing, or catering, has become one of the preconditions for successful market competitiveness and enable airlines to achieve efficiency and increase service ability (Dožić 2019). An

[11] https://www.easa.europa.eu/sites/default/files/dfu/Weight%20Survey%20R20090095%20Final.pdf

airline must perform all activities to maintain the aircraft in a condition that is in accordance with the standardized maintenance. In general, the different activities, also known as Maintenance, Repair, Overhaul (MRO), are performed in different places (the platform, hangars, or the technical bases), where facilities and personnel for certain checks are provided (Schmitt and Gollnick 2016). The following activities refer to maintenance and repair, and they include:

- inspections that cover checks performed before and after each flight (performed during turnaround time by the cockpit and ground crews), and they are related to checks of aircraft and systems status and functions.
- services that encompass refill of lubricants, cleaning, and intensified functional checks of the aircraft and its systems by the airline ground station maintenance crew in a hangar, typically overnight on a regular time base. Additionally, services include the replacement of defective parts or components.
- repair is dedicated to replacement or overhaul of the defective parts or components, performed on or off the aircraft, by the maintenance organization.
- modification and improvement of components refers to small adaptions of components and parts and software, usually performed due to safety, functionality or performance improvement, at the request of manufacturers and certification authority.

On the other hand, all the above-mentioned activities, performed to recover the safe of operational state of the aircraft and engine systems, components and parts, are considered an overhaul (Schmitt and Gollnick 2016).

While providing corresponding maintenance activities for the fleet, the airline should meet a number of maintenance efficiency goals (Wensveen 2015). The out-of-service time of each aircraft should be minimized, the utilization should be maximized, and personnel utilization should be optimized, but the required level of safety also needs to be maintained.

It should be noted that airlines develop their own maintenance program, which must comply with the recommendations and requirements requested by manufacturers and regulatory bodies. There are four types of aircraft maintenance: routine scheduled maintenance (airframe and engine checks), nonroutine maintenance (structural fatigue repair and corrosion control, performed as a result of unforeseen events, such as an accident or accidental event, or in response to an airworthiness directive), refurbishments (cabin upgrades and exterior painting), and modifications (passenger-to-freight conversions and hush kit installations) (Wensveen 2015). Routine maintenance of airframe includes:

- visual inspection, which implies that pilots and mechanics walk around the aircraft before the flight to ensure that there are no obvious problems such as leaks, missing rivets, or cracks.
- overnight maintenance, which is performed at the end of each working day. It is performed by workers who spend 1 to 1½ hours to ensure that the aircraft is operating in accordance with the original equipment manufacturers (OEM's) minimum equipment list (MEL).
- the A-check, which is cyclically conducted every 125 flight hours (approximately two or three weeks) and it includes a more detailed visual inspection of the fuselage exterior, engine, and certain readily accessible subsystems, avionics, and accessories.
- the B-check, which is performed every 750 flight hours (three to four months) by workers who conduct an open inspection of panels and cowlings, during which some preventive maintenance (exterior wash, engine oil spectroscopic analysis, etc.) is provided, oil filters are checked and changed, parts are lubricated, and the airframe is carefully examined.
- the C-check is carried out approximately every 3,000 flight hours or every 15 months, and it incorporates both the A and B checks. In addition, it includes reparation of components, calibration of flight controls, and the testing of major internal mechanisms, as well as heavy lubrication, attendance to Service Bulletin requirements, minor structural inspections, flight control rigging tests, engine baroscopic inspections, compressor washes, aircraft appearance maintenance, some corrosion prevention, and a post-check flight test.
- the D-check is the most comprehensive form of routine maintenance and it is carried out approximately every six to eight years, or approximately every 20,000 flight hours. During this check seats, galleys, lavatories, cockpit, furnishings, headliners, and sidewalls are removed from the passenger cabin to enable careful structural inspections. Moreover, the aircraft is rebuilt in order to return to its original condition as much as possible.

Overnight maintenance and checks A and B are usually performed during periods without planned traffic (overnight) to avoid interruption of planning activities, and they can be performed at the airport or even at the ramp. Checks C and D are more complex and require special equipment. Moreover, the aircraft needs to be out of service for a longer time. Table 2.2 provides location and duration of routine maintenance activities (Remzi et al. 2016).

Table 2.2. Typical maintenance checks.

Check	Location	Duration
Visual inspections/overnight	At gate	~ 1 hour
A	At gate	~ 10 hours
B	At gate	~ 10 hours – 1 day
C	Hangar	~ 3 days – 1 week
D	Hangar	~ 1 month

2.8 Aircraft classification

Airlines use different types of aircraft when providing civil air transport. Aircraft types differ in characteristics and performance. To facilitate the exchange of information and aircraft operations in air transport, the need appeared to provide a suitable classification of aircraft.

Depending on the construction of the aircraft, aircraft heavier than air (aerodyne) and aircraft lighter than air (aerostats) can be distinguished (Figure 2.19). In the case of lighter-than-air aircraft, their ability to climb depends on how much the aircraft is filled with the lighter-than-air gas. The aircraft can be filled with hot air, helium, or hydrogen. Balloons and airships belong to the category of lighter-than-air aircraft.

In the case of heavier-than-air aircraft, the lift force is generated by the flow of air over an aerofoil. Depending on the propulsion they use to move through the air, heavier-than-air aircraft are divided into two categories: unpowered and powered aircraft. Unpowered aircraft are vehicles that

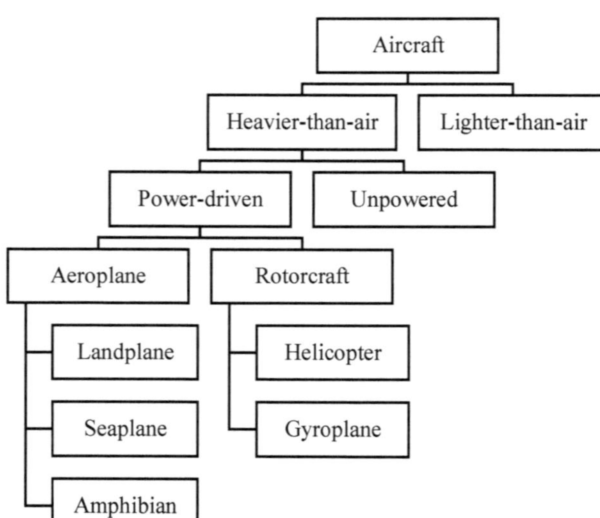

Figure 2.19. Aircraft classification.

can fly without any propulsion mechanism. These vehicles move through the air using aerodynamic forces. The most common representatives are gliders. Powered aircraft use engine power to move through the air. They can further be divided into rotorcrafts and aeroplane. According to ICAO (2018), a rotorcraft is a power-driven heavier-than-air aircraft supported in flight by the reactions of the air on one or more rotors. Helicopters and gyroplanes are examples of rotorcrafts. Aeroplane is a power-driven heavier-than-air aircraft, deriving its lift in flight chiefly from aerodynamic reactions on surfaces (ICAO 2018).

Depending on the surface they use to take-off and land, airplanes can be divided into landplanes (they have landing gear), seaplanes (they have floats), and amphibians (which can use both land and water to land and take-off).

The ICAO's document *DOC 8643—Aircraft Type Designators*, contains designators for those aircraft types that are most commonly provided with air traffic service (ATS). This document consists of four parts which provide: Part 1—the common names of the aircraft manufacturers, followed by the aircraft types they produce(d), Part 2—the aircraft types by their type designator in alpha-numeric order, Part 3—the aircraft types by their model number and/or model name, and Part 4—the common names of the aircraft manufacturers followed by their full name(s). ICAO assigns an aircraft type designator to all aircraft heavier than ultra-light. Those aircraft types with a maximum certified take-off mass of 454 kg and a stalling speed not greater than 65 km/h (35 kt) are considered ultra-light aircraft. There are also ultra-light helicopters and gyrocopters, and they have a maximum certified take-off mass of 454 kg, (ICAO 2016). A designator for aircraft type consists of not more than four characters. The primarily intention of designators is their use in flight plans and associated air traffic services messages, for the purpose of facilitating data exchange, increasing automation, and avoiding misunderstanding.

A designator is derived from the manufacturer's model number or model name (Table 2.3). In addition to the aircraft type designators, ICAO also defines the description of the aircraft type using three characters. Each of the characters corresponds to the specific characteristics of the aircraft (ICAO 2016):

- The first character is dedicated to the aircraft category:
  - L – *Landplane,*
  - S – *Seaplane,*
  - A – *Amphibian,*
  - H – *Helicopter,*
  - G – *Gyrocopter,*
  - T – *Tilt-wing aircraft,*

Table 2.3. Aircraft type designator and description of aircraft type.

Aircraft type	ICAO aircraft type designator	ICAO description of the aircraft type	WTC
Airbus A-320neo	A20N	L2J (*landplane, 2 jet engines*)	M
Airbus A-380-800	A388	L4J (*landplane, 4 jet engines*)	H
Boeing 737-800	B738	L2J (*landplane, 2 jet engines*)	M
Boeing 747-8	B748	L4J (*landplane, 4 jet engines*)	H
Canadair Cl-215	CL2P	A2P (*amphibian, 2 piston engines*)	M
ATR 72-600	AT76	L2T (*landplane, 2 turboprop engines*)	M
Cessna 172	C172	L1P (*landplane, 1 piston engine*)	L
Black Hawk	H60	H2T (*helicopter, 2 turboprop engines*)	-

- The second character refers to the number of aircraft engines; therefore, it is a number or C in the case of fixed-wing aircraft only and it indicates that two engines are coupled to drive a single propeller system;
- The third character describes the type of engine:
  - o P – *piston,*
  - o T – *turboprop,* and
  - o J – *jet.*

The last column of Table 2.3 provides the wake turbulence category (WTC):

- L – *light,*
- M – *medium,*
- H – *heavy.*

The aircraft manufacturer provides appropriate specifications containing information for each type of aircraft, such as general aircraft characteristics, avionics, engines, performance, noise levels, emissions, dimensions, and masses.

General aircraft characteristics include data related to the number of cabin and flight crew required, passenger cabin capacity (number of seats), and available cargo space. The flight crew encompasses pilots and co-pilots who are in charge of operating the aircraft. Cabin staffs are stewards and flight attendants who are available to assist and help passengers during the flight. The required number of cabin and flight crew members depends on the type of aircraft, its capacity, and the number of doors on the aircraft.

The capacity of the aircraft represents the number of passenger seats offered in the passenger cabin. According to the capacity they offer, aircraft can be divided into:

- Very large wide-bodies (e.g., Airbus A380, Airbus A340, Boeing B747),
- Wide-bodies (e.g., the A330 family: A330-200, A330-300, A330-800neo, A330-900neo; the A350 family: A350-800, A350-900, A350-1000; B777, B787 (B787-8, B787-9, B787-10)),
- Narrow-bodies (e.g., the A320 family: A318/319/320/321, A318neo/319neo/320neo, 321neo; B737 NG, B737 max), and
- Regional (e.g., ATRs, CRJs, ERJs).

The capacity differs depending on the aircraft type. Moreover, the capacity can differ even for the same aircraft type. Depending on whether the airline wants to have different passenger classes in its offer, it chooses the appropriate passenger cabin configuration, where the number of offered passenger seats can vary. Seats, as well as the seat pitch, usually differ in different passenger classes. The seat pitch represents the distance between two identical points on two seats located one behind the other along the longitudinal axis of the aircraft. It also represents one of the parameters of comfort in the cabin. Another parameter that refers to comfort is legroom—the space available for passengers to stretch their legs while seated (Figure 2.20). Figures 2.21 and 2.22 show a longitudinal section of an Airbus 320 in two different configurations (a single passenger class configuration in Figure 2.21 and a configuration in which there are two passenger classes in Figure 2.22).

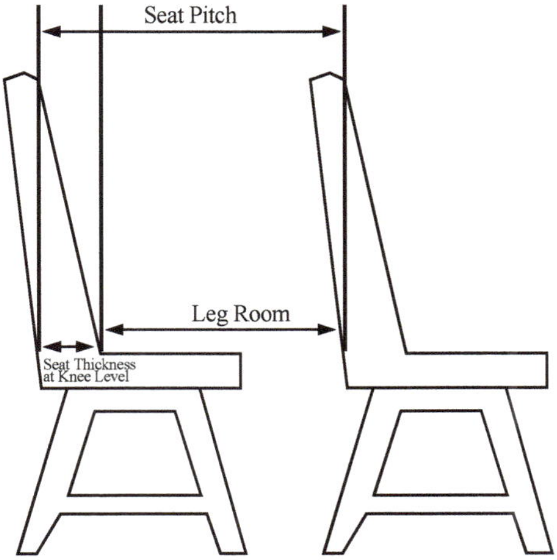

Figure 2.20. Seat pitch and legroom.

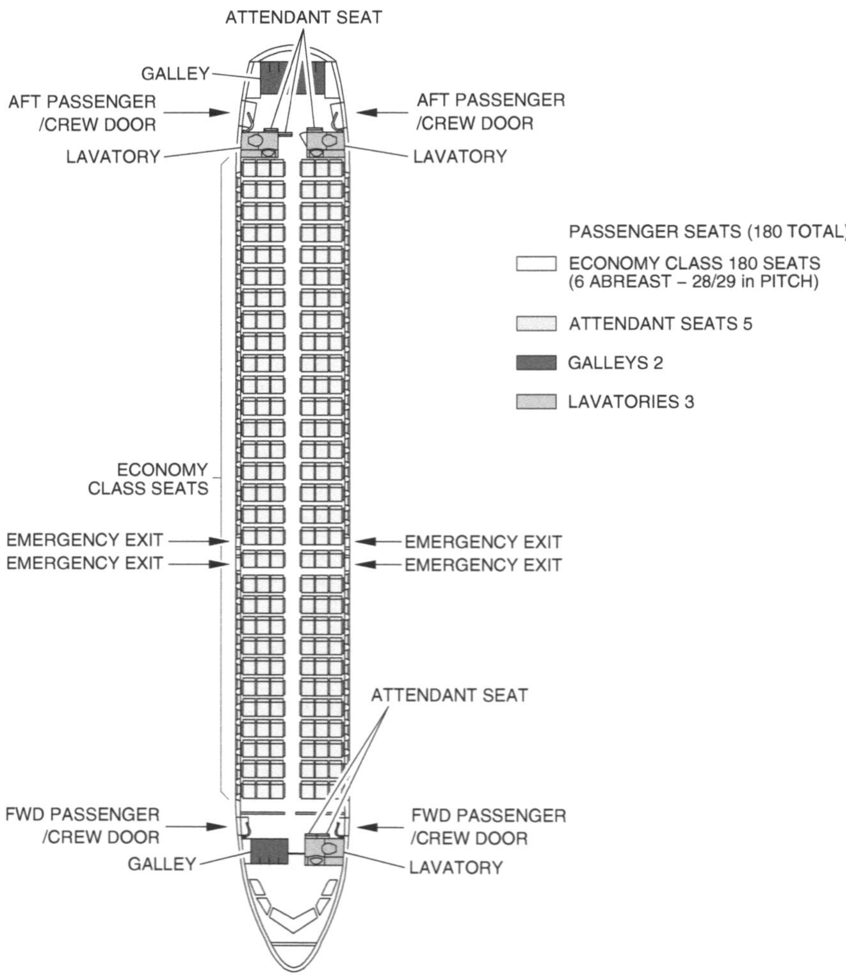

Figure 2.21. Typical configuration of A320neo—single-class, high density.
Source: Airbus 2005

The available cargo space in the aircraft itself depends on the purpose of the aircraft, i.e., on whether the aircraft transports passengers and goods, or goods only.

Aircraft performance includes data related to the range of the aircraft, take-off and landing distances, different aircraft speeds, payload-range diagrams, runway requirements, and noise performance. These factors could limit aircraft operations at a particular airport (Dožić and Kalić 2018). For example, aircraft range defines the maximum distance between take-off and landing which could be flown without refuelling. Depending

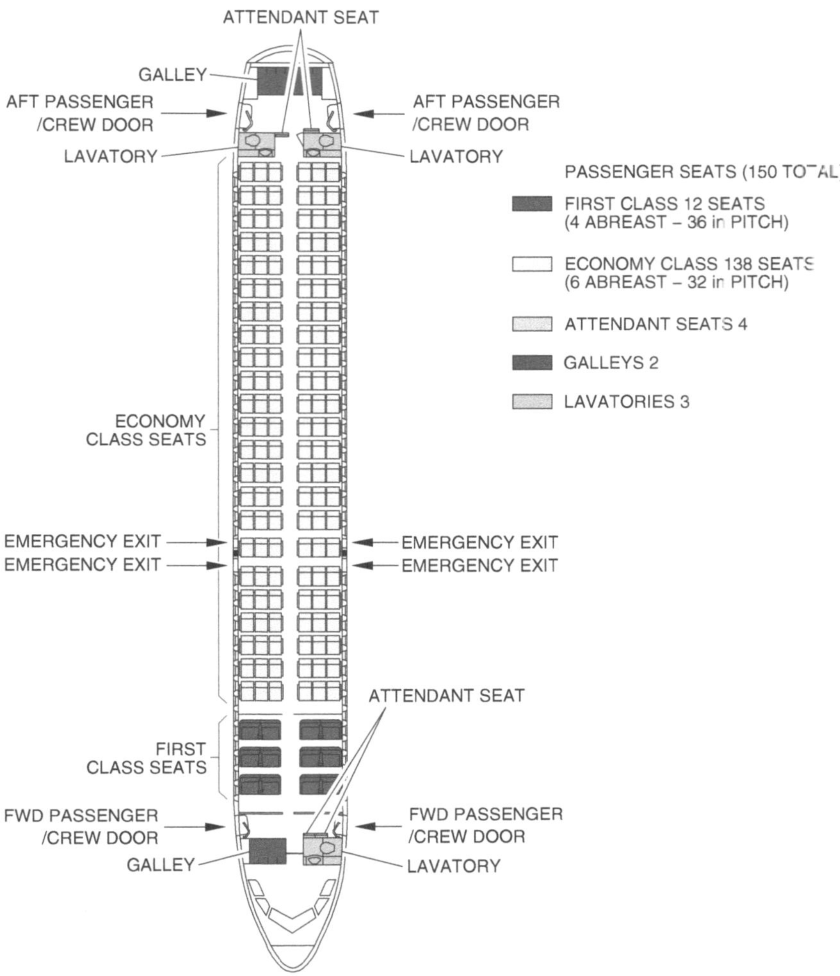

Figure 2.22. Typical configuration of A320neo—two-class.
Source: Airbus 2005

on the aircraft range, the set of corresponding airports that could be served from the considered airport is defined (Figure 2.23).

The manufacturer also provides the appropriate dimensions of the aircraft. Usually, the three projections of an aircraft are provided (Figure 2.24). Aircraft dimensions are very important due to the use of airport infrastructure and equipment. Moreover, the defined standards related to the separation of aircraft on the runway, taxiways, manoeuvring areas and platforms need to be satisfied to achieve the appropriate level of safety.

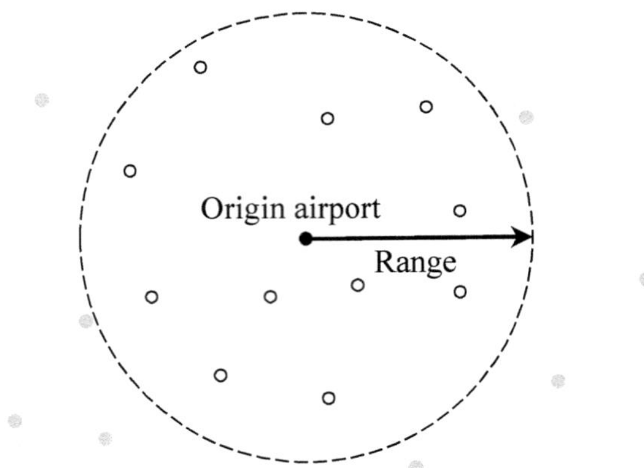

Figure 2.23. Aircraft range and airport catchment area.

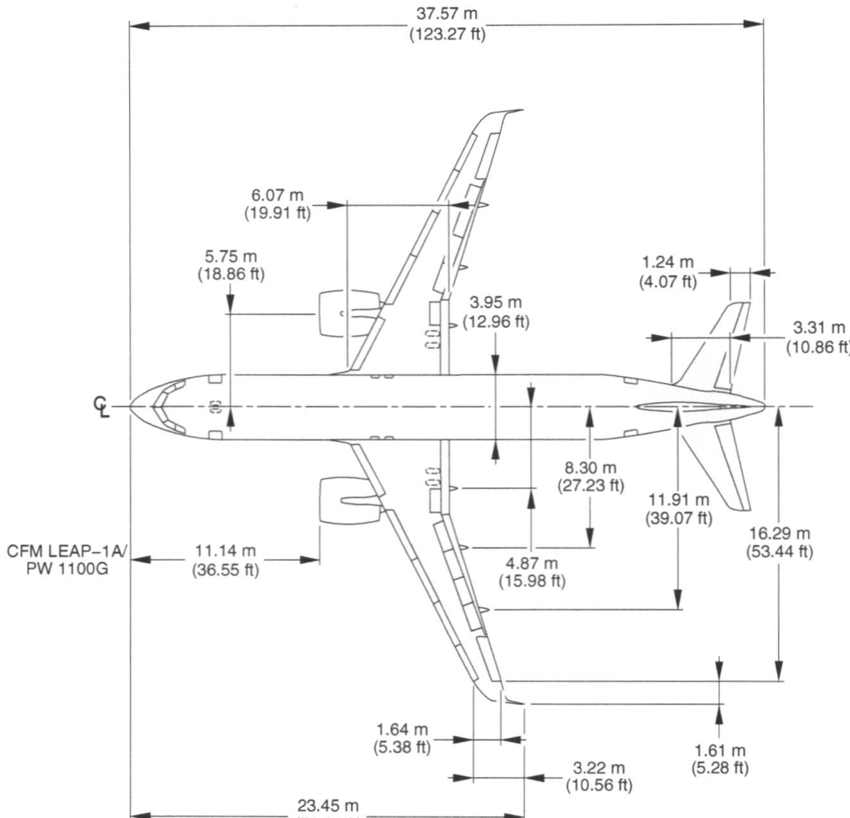

Figure 2.24 contd. ...

...Figure 2.24 contd.

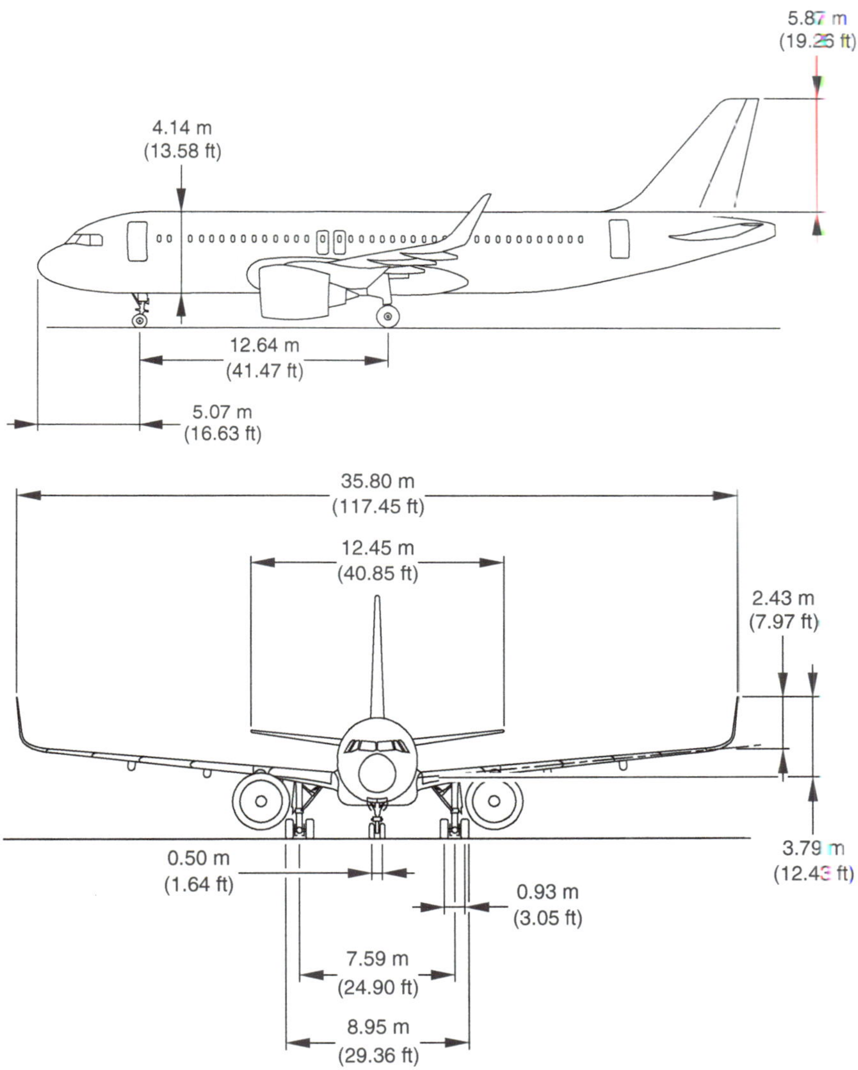

Figure 2.24. General aircraft dimensions—Airbus A320neo.
Source: Airbus 2005

The cross-sections and longitudinal sections of the aircraft provide the necessary parameters related to the comfort in the passenger cabin (in the appropriate passenger classes), Figures 2.25 and 2.26.

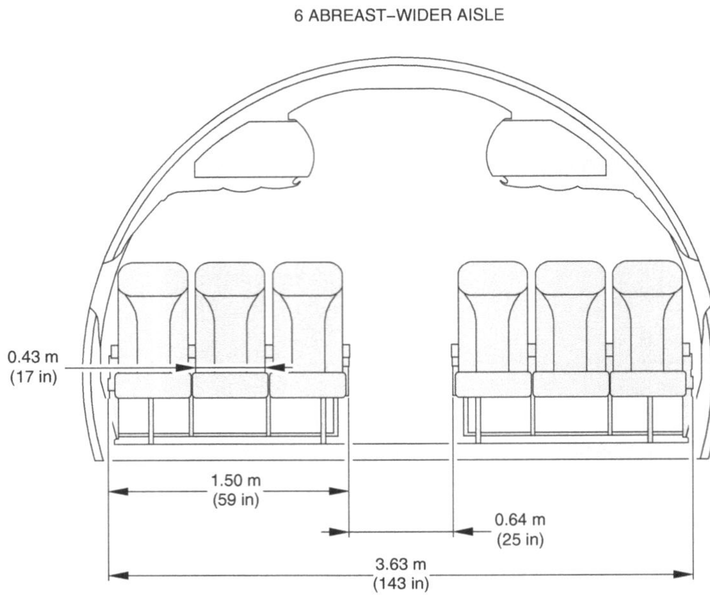

Figure 2.25. Cross section of A320neo—economy class.
Source: Airbus 2005

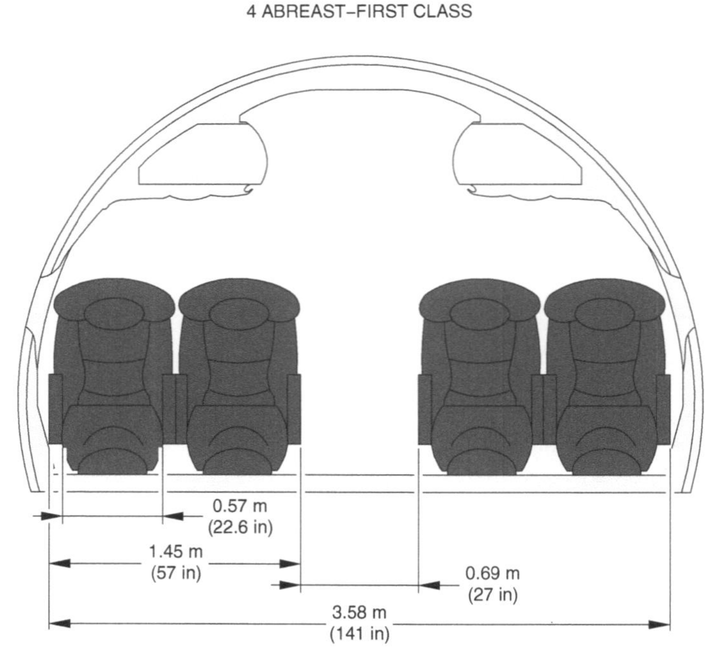

Figure 2.26. Cross section of A320neo—first class.
Source: Airbus 2005

References

Airbus. 2002. Getting to Grips With Aircraft Performance. Flight Operations Support & Line Assistance, January 2002.

Airbus. 2005. A320 Aircraft Characteristics, Airport and Maintenance Planning. Airbus S.A.S, France.

Boyer, R. and Padmapriya, N. 2016. Aircraft Materials. Reference Module in Materials Science and Materials Engineering. doi: 10.1016/b978-0-12-803581-8.01934-2.

Dožić, S. and Kalić, M. 2018. Aircraft type selection problem: application of different MCDM methods, the advances in intelligent systems and computing series, advanced concepts. Methodologies and Technologies for Transportation and Logistics 572: 156–175.

Dožić, S. 2019. Multi-criteria decision making methods: application in the aviation industry. Journal of Air Transport Management 79, 101683.

FAA. 2016a. Pilot's Handbook of Aeronautical Knowledge, FAA-H-8083-25B – Chapter 3, Aircraft construction, 3-4. https://www.faa.gov/regulations_policies/handbooks_manuals/aviation/phak/.

FAA. 2016b. Pilot's Handbook of Aeronautical Knowledge, FAA-H-8083-25B - Chapter 4, Principles of Flight, 4-6. https://www.faa.gov/regulations_policies/handbooks_manuals/aviation/phak/.

FAA. 2016c. Pilot's Handbook of Aeronautical Knowledge, FAA-H-8083-25B - Chapter 6.

FAA. 2016d. Pilot's Handbook of Aeronautical Knowledge, FAA-H-8083-25B - Chapter 7, Aircraft Systems.

Fioriti, M., Boggero, L., Prakasha, P.S., Mirzoyan, A., Aigner, B. and Anisimov, K. 2020. Multidisciplinary aircraft integration within a collaborative and distributed design framework using the AGILE paradigm. Progress in Aerospace Sciences, Volume 119, November 2020, 100648.

https://www.faa.gov/regulations_policies/handbooks_manuals/aviation/phak/media/08_phak_ch6.pdf.

https://www.faa.gov/regulations_policies/handbooks_manuals/aviation/phak/media/09_phak_ch7.pdf.

ICAO. 2009. ANNEX References for Ramp Inspections Guidance, https://www.icao.int/safety/fsix/Library/Reference_ANNEXES_for_ICAO_Ramp_Inspections_Guidance_2009_07_02.pdf.

ICAO. 2016. DOC 8643 – Aircraft Type Designators.

ICAO. 2018. International Standards and Recommended Practices: Annex 1 to the Convention on International Civil Aviation, Personal Licensing, 12th edition.

Isermann, R. 2006. Reliability, Availability and Maintainability (RAM). *In*: Fault-Diagnosis Systems. Springer, Berlin, Heidelberg. https://doi.org/10.1007/3-540-30368-5_3.

Johansson, C., Derelov, M. and Olvander, J. 2017. How to use an optimization-based method capable of balancing safety, reliability, and weight in an aircraft design process. Nuclear Engineering and Technology 49: 404–410.

McCormick, B.W. 1995. Aerodynamics, Aeronautics, and Flight Mechanics. Second edition. John Wiley & Sons.

Modern Airliners. What are the Specs of the Airbus A380 Super Jumbo? https://modernairliners.com/airbus-a380/airbus-a380-specs/.

O'Connor, P.D.T. and Kleyner, A. 2012. Practical Reliability Engineering, Fifth Edition, Wiley, A John Wiley & Sons, Ltd., Publication.

Razumenić, S. 1984. Performanse Transportnih Vazduhoplova, Saobraćajni Fakultet, Beograd.

Saltoğlu, R., Humaira, N. and Enalhan, G. 2016. Aircraft scheduled airframe maintenance and downtime integrated cost model. Advances in Operations Research, Volume 2016, Article ID 2576825.

Schmitt, D. and Gollnick, V. 2016. Air Transport System, Springer Wien Heidelberg New York Dordrecht London.

Taneja, N. 1989. Introduction to Civil Aviation. Second edition. Lexington Books. D.C. Heath and Company/Lexington, Massachusetts/Toronto.

Wensveen, J. 2015. Air Transportation: A Management Perspective. Publisher: Routledge; 8th edition.

Chapter 3
Airports

3.1 Introduction

Airports are one of the three main parts of the air transport system. As mentioned before, airports, airlines, and ATC/ATM (Air Traffic Control/Management) are subsystems of the core air transport system. Airports and ATC/ATM represent infrastructure components. According to the ICAO[1] definition, an aerodrome is a defined area on land or water (including any buildings, installations, and equipment) intended to be used either wholly or in part for the arrival, departure, and surface movement of aircraft. This chapter is dedicated to commercial airports, which are aerodromes with extended facilities. The users of airports are airlines, their aircraft, and passengers, as well as freight/cargo shipments and mail. As ground-based infrastructure, airports provide space, facilities, and equipment for serving users. Additionally, airports provide the physical connection between air and different ground transport modes.

Every airport consists of the landside and the airside area. The demand in the landside area consists of passengers and cargo shipments, while in an airside area it consists of air transport movements. An air transport movement or operation is a landing or take-off of an aircraft operating a scheduled (any flight with a scheduled timetable) or non-scheduled service (e.g., charter).

Generally, air travels are intermodal journeys because, in addition to airplanes, passengers always use different transport modes from their origin to the airport and from the airport to their final destination. For air passengers, the airport is a central multimodal transport node, connected to the entire transport system (road, rail, and waterway). Air passengers

[1] ICAO Annex 14, Doc. 9157, Aerodrome Design Manual, Part 1, Runways, 4th edition, 2020.

envision a smooth and safe door to door trip, in which the air transport mode plays a dominant role (Janić 2013). This implies efficient and effective airport accessibility by ground transport modes, fast processing through the terminal, and flights without delays. The airport has a key role in providing not only physical capacity (facilities and equipment) but also effective management, organization, and service delivery.

This chapter examines commercial air transport airports, with scheduled and non-scheduled traffic and includes considerations of the world's busiest airport systems by passenger count, airport codes, airport classification, the airport access system and its catchment area, the airport complex and operations, airport demand and capacity, and airport business and economics.

3.2 The busiest airports in the world

As part of the air transport system, airports serve passengers, cargo and aircraft, to carry out air transport between various destinations. Airports can be of various sizes, depending on the number of passengers and aircraft that they handle. We will first look at airports from the standpoint of the number of passengers that they handle during a year. The total number of passengers (PN) at a given airport is obtained by adding up the number of passengers who started their travel (a), ended their travel (b), disembarked from the aircraft, waited for their connected flight at the airport, and continued their trip on another aircraft (c), or who were in transit (d).[2] The third category of passengers (c) is transfer passengers, where the time that they spend at the airport, i.e., between the two flights, is less than 24 hours. Transfer passengers are counted twice, when exiting the aircraft and when entering the aircraft. Transit passengers are counted once because they stop at the airport briefly and continue their journey on the same aircraft on which they arrived (the aircraft with the same registration number and the same flight number). Therefore, the total number of airport passengers is calculated based on the following formula:

$$PN = (a + c) + (b + c) + d$$

where (a+c) represents the total number of embarked passengers, while (b+c) represents the total number of disembarked passengers.

Airports can also be ranked based on the number of operations (air transport movements). The number of operations is the sum of the number of take-offs and the number of landings. The number of passengers and the number of operations are two different criteria, therefore it is possible for certain airports to have the same or a similar number of operations,

[2] ACI Statistic Manual, 2011.

and to differ significantly in the number of passengers: for example, Beijing Capital International Airport (China) and Denver International Airport (USA) have a similar number of operations (close to 600,000 annually), while the number of passengers differs significantly (around 100 million compared to slightly more than 60 million, respectively). This can be explained by the fact that higher capacity aircraft service the Beijing airport. Also, Hartsfield-Jackson Atlanta International Airport (USA) and Beijing Capital International Airport, which both have more than 100 million passengers per year, differ greatly in the number of operations (Atlanta is ranked top in the world, both according to the number of passengers and according to the number of operations, while the Beijing airport is second in the world according to the number of passengers, and sixth according to the number of operations).

Figure 3.1 depicts the number of airports according to the number of passengers, by interval, from the list of the 50 busiest airports in the world, by passenger traffic. Only two airports handle more than 100 million passengers (Atlanta and Beijing), i.e., 27 airports handle more than 50 million passengers per year. Of the top 50 airports in the world by passenger traffic, as many as 15 are in the United States of America. These are Atlanta (ATL), Los Angeles (LAX), Chicago (ORD), Dallas (DFW), Denver (DEN), New York (JFK), San Francisco (SFO), Seattle (SEA), Las Vegas (LAS), Orlando - Florida (MCO), Charlotte - North Carolina (CLT), Newark (EWR), Miami (MIA), Phoenix (PHX), and Houston (IAH). Also, the list of the 50 busiest airports includes eleven airports in Europe.

For the past 21 years, the Hartsfield-Jackson Atlanta International Airport has been the top ranking airport in the world. Considering that this airport handles more than 107 million passengers annually (2018), it is an impressive fact that the daily number of passengers is around 300,000.

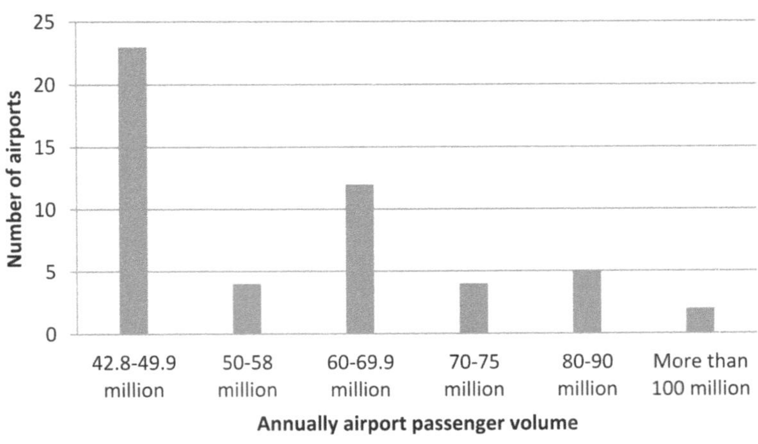

Figure 3.1. Number of airports by annual passenger volume in 2018.

Also, this airport conducts around 900,000 take-off and landing operations annually, and loads and unloads close to 700,000 tonnes of cargo.

The Hartsfield-Jackson Atlanta International Airport reached 100 million passengers in 2015 (it had 78 million passengers in 2000). From 2015 to 2018 the airports in Atlanta, Beijing and Dubai held the top three places, respectively, according to the number of passengers. As shown in Table 3.1, of the top ten airports in the world, three are located in the United States of America, five are in Asia (China, Hong Kong Special Administrative Region, Japan, UAE), and two are in Europe.

The list of European airports that had more than 40 million passengers annually (according to Eurostat data from 2018) includes London Heathrow (LHR), Paris Charles de Gaulle (CDG), Amsterdam Schiphol (AMS), Frankfurt (FRA), Istanbul Atatürk (IST), since April 2019 Istanbul Havalimanı (the entire passenger traffic was transferred),

Table 3.1. The top ten airports in the world.[3]

Rank	Airport	Country	Name of the airport	IATA code	Number of passengers in 2018 (million)
1.	Atlanta	USA	Hartsfield–Jackson Atlanta International Airport	ATL	107.4
2.	Beijing	China	Beijing Capital International Airport	PEK	101.0
3.	Dubai	United Arab Emirates	Dubai International Airport	DXB	89.1
4.	Los Angeles	USA	Los Angeles International Airport	LAX	87.5
5.	Tokyo	Japan	Tokyo Haneda Airport	HND	87.1
6.	Chicago	USA	O'Hare International Airport	ORD	83.3
7.	London	United Kingdom	London Heathrow	LHR	80.1
8.	Hong Kong	Special Administrative Region of China	Hong Kong International Airport (Chek Lap Kok Airport)	HKG	74.5
9.	Shanghai	China	Shanghai Pudong International Airport	PVG	74.0
10.	Paris	France	Paris Charles de Gaulle Airport	CDG	72.2

[3] https://edition.cnn.com/travel/article/worlds-busiest-airports-2018/index.html

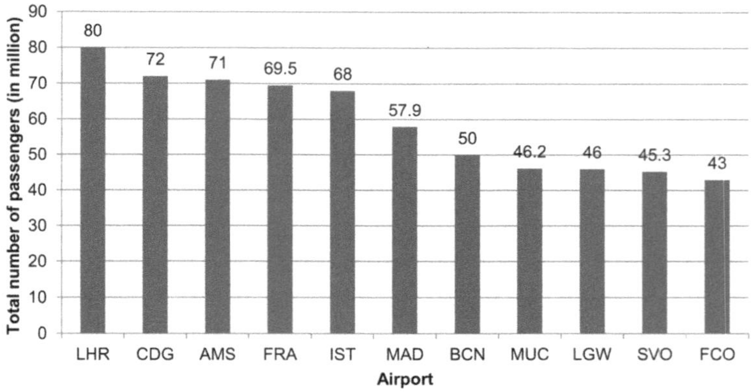

Figure 3.2. Leading airports in Europe according to the total number of passengers (in million).[4]

Madrid Barajas (MAD), Barcelona El Prat (BCN), London Gatwick (LGW), Moscow Sheremetyevo (SVO), and Rome Fiumicino (FCO) (Figure 3.2). These airports can be said to be very large. The group of large European airports, with between 25 and 40 million passengers annually, consists of 13 airports, including Paris Orly, Moscow Domodedovo, as well as the airports in Zurich, Copenhagen, Stockholm, and Vienna.

Certain cities have multiple airports (Figure 3.3). For example, London has as many as six airports: London Heathrow (80 million passengers), London Gatwick (46 million), London Stansted (28 million), London

Figure 3.3. Airports in London.

[4] Eurostat, Press Release, 186/2019 - 6 December 2019, Eurostat Press Office.

Luton (16.8 million), London City (4.8 million), and London Southend (1.5 million). Therefore, the total number of passengers at airports in London in 2018 was around 97 million. Paris has three airports: Charles de Gaulle (72.2 million passengers), Orly (33 million), and Beauvais (3.8 million), with a total of around 109 million passengers handled by these airports. Also, Milan has three airports: Malpensa (24.7 million passengers), Linate (9.2 million), and Caravaggio-Bergamo (13 million). These three airports handle close to 47 million passengers annually.[5]

Air transport is most developed in the United States of America. This is to be expected, bearing in mind that the USA has a population of more than 320 million and a territory that is the size of a continent. Specifically, the USA has an area of 9.8 million km^2, while all of Europe has an area of 10.5 million km^2, with the territory of the European Union being 4.4 million km^2. Europe has a population of 740 million.[6]

In the past ten years, China has recorded the fastest air transport growth rate. In 2000 airlines registered in China carried close to 62 million passengers, while airlines in the USA carried more than 660 million passengers. Between 2009 and 2014 the number of passengers in China increased more than 47%, from 266 million to 391 million. In 2017 airlines registered in China carried more than 550 million passengers, ranking second, behind airlines registered in the USA, which carried 850 million passengers. Comparing the number of passengers in 2000 and 2018, it is apparent that the number of passengers in the US increased by around 35%, while the number of passengers in China increased tenfold.

China, the USA, and the European Union are among the world's leading economic powers. Tables 3.2 and 3.3 present their basic socioeconomic characteristics, as well as indicators that are related to air transport. Nearly one-third of the world's population lives in the European Union member states, China, and the United States of America. China is the third or fourth (depending on the source) largest country in the world, after Russia and Canada, with an area of more than 9.5 million km^2, approximately the same as the USA. Also, China has the largest population in the world, around 1.4 billion.

This group of economic powers could also include Japan, considering its economic development and the fact that the Tokyo Haneda airport is among the top ten in the world. In addition to Tokyo Haneda (87.1 million passengers), the capital of Japan also has two more airports: Narita (42.6 million passengers) and Ibaraki (760,000 passengers). The three airports handled 130.5 million passengers, combined, in 2018. According to World Bank data for 2018, Japan has a population of 126.5 million, a GDP of $ 4,971 billion, and GNI per capita of $ 41,340. Regarding air

5 All data are taken from airport websites.
6 Data source: data.worldbank.org

Table 3.2. Demographic and economic characteristics of China, the European Union, and USA (2018).

Country or state union	Population (million)	Area (million km²)	Percentage of global population	GDP (billion $)	GNI per capita (Atlas method)
China	1,390	9.56	18.6%	13,608	9,470
European Union	513	4.38	6.6%	18,756	35,309
USA	327	9.83	4.3%	20,494	62,850

Sources: https://data.worldbank.org, https://www.worldometers.info/population/china-eu-usa-japan-comparison/

Table 3.3. Tonne-kilometres and number of passengers in China, the European Union, and the USA (2018).

Country or state union	Air transport cargo carriage (million tonne-kilometres)	Number of passengers carried by airlines registered in the given country (millions)
China	25,256	611
European Union	37,929	802
USA	42,985	889

Source: https://data.worldbank.org

transport indicators, Japan had 9,420 million tonne-kilometres of cargo and 126 million passengers carried by airlines registered in this country.

3.3 Airport codes

Airport codes are of great importance in the process of transport planning, operative traffic handling, and execution and monitoring of the transport of passengers, their luggage, cargo, and mail. Some cities have multiple airports in their vicinity, which is why the airports must have codes so that there would be no confusion or mistakes. Every airport has two designations: an ICAO code and an IATA code.[7]

The ICAO code represents a location indicator. It is a four-letter designation: the first letter indicates the region of the world, the second the country, and the third and fourth indicate the airport. This code is used to identify the airport location, which is why it is used in air traffic control and flight planning. Figure 3.4 shows the regions of the world according to the first letter of the ICAO code.

[7] https://www.world-airport-codes.com/; https://www.icao.int/wacaf/documents/apirg/sg/2012/afi_opmet_mtf4/docs/wp10_atta.pdf

Europe is divided into two regions: the northern part uses the letter E and the southern uses the letter L. Table 3.4 shows the codes of certain airports in Europe. Countries are designated by one letter. For example, France uses the letter F, therefore the ICAO codes for all airports in France are LF**. Both Germany and Croatia use the letter D, but the ICAO codes

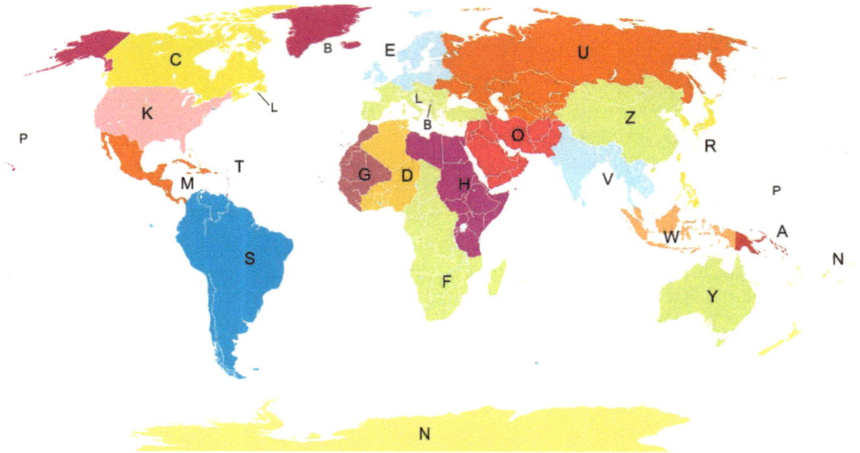

Figure 3.4. Regions of the world and the first letter of the ICAO code.

Table 3.4. ICAO and IATA codes of European airports.

Airport	Country	ICAO code	IATA code
Brussels Airport	Belgium	EBBR	BRU
Munich Airport	Germany	EDDM	MUC
Copenhagen Airport, Kastrup	Denmark	EKCH	CPH
Stockholm Arlanda Airport	Sweden	ESSA	ARN
Oslo Airport, Gardermoen	Norway	ENGM	OSL
Amsterdam Airport Schiphol	The Netherlands	EHAM	AMS
Luxembourg Airport, Findel	Luxembourg	ELLX	LUX
Warsaw Chopin Airport	Poland	EPWA	WAW
Athens Int'l. Airport Eleftherios Venizelos	Greece	LGAV	ATH
Vienna International Airport	Austria	LOWW	VIE
Rome–Fiumicino Int'l. Airport Leonardo da Vinci	Italy	LIRF	FCO
Adolfo Suárez Madrid–Barajas Airport	Spain	LEMD	MAD
Budapest Ferenc Liszt International Airport	Hungary	LHBP	BUD
Belgrade Nikola Tesla Airport	Serbia	LYBE	BEG

for Germany's airports are ED**, while the codes for Croatia's airports are LD**. The Dominican Republic also uses the letter D, but the first letter of the ICAO code is M, since it is in the Caribbean.

In the case of the USA, the first letter indicates the country, and the second third, and fourth are the airport's IATA code. The letter K covers the 48 contiguous USA states (not including Alaska and Hawaii). For example, KIAD is the Dulles International Airport, in Washington D.C., while KORD designates O'Hare International Airport in Chicago. Similarly, Canada also uses a single one-letter designation, and the ICAO code consists of the letter C and the airport's IATA code.

Unlike the ICAO code, which represents a location indicator, the airport's IATA code is used for tagging luggage and it is given on the e-ticket itinerary. This is a three-letter code, commonly generated based on the name of the city. For example, the IATA code of the Hartsfield-Jackson Atlanta International Airport is ATL, Denver International Airport is DEN, San Francisco International Airport is SFO, Phoenix Sky Harbor International Airport is PHX, etc.

3.4 Airport classification according to size and type of traffic

There are different classifications of airports into categories. Classification can be done based on the number of passengers, the quantity of cargo that is handled by the airport, according to the type of traffic (intercontinental, domestic, origin-destination), according to certain physical characteristics (e.g., number of gates), according to their connection to other airports in the network, or according to their main role (leisure, low-cost carrier, charter, regional, local).

It is commonly accepted that very small airports are those with less than a million passengers per year. The larger category, with between 1 and 5 million passengers annually, is considered small airports. Airports with between 5 and 15 million passengers, and between 15 and 25 million passengers, are medium-size. Large airports handle between 25 and 40 million passengers, and very large airports have more than 40 million passengers per year.

Humphreys and Ison (2005) classified airports in the United Kingdom into three categories, according to the number of passengers handled: large airports with more than 10 million passengers, medium airports with between 2 and 10 million passengers, and small—with fewer than 2 million passengers. The United Kingdom has a network of as many as 65 airports, of which two are very large (London Heathrow and London Gatwick, with 80 million and 46 million passengers in 2019, respectively), two large airports (Manchester and London Stansted, with around

29 million and 28 million passengers, respectively), followed by London Luton, with 18 million passengers. The group of airports with between 5 and 15 million passengers consists of Edinburgh, Birmingham, Bristol, Glasgow, Belfast International, Newcastle International, London City, and Liverpool. A total of 41 airports in the network have more than 100,000 passengers per year. Of the 24 airports that are public service obligation routes, as many as 18 airports are either very small or are not included in the published statistics.

According to their roles, airports can be classified as primary hubs (major hubs, mega hubs), secondary hubs, regional airports, and tourist airports (leisure). Hubs represent network nodes that are directly linked to a large number of nodes (destinations) and through which transit and transfer traffic is conducted. Hubs are linked to feeder airports (smaller hubs, secondary airports, regional airports, etc.) therefore passengers congregate in the hub, from where they continue their travel. Therefore, a feeder airport can be defined as an airport supporting a hub.

Primary hubs include very large and large airports where a great number of destinations are offered, domestic and international traffic is conducted, both full-service carriers (FSC) and low-cost carriers (LCC) are present, with a large number of transit and transfer passengers. Secondary hubs are airports that may have all the characteristics of primary hubs, with the difference being the number of passengers and operations. These are usually mid-sized airports. Secondary hubs are feeders for primary (mega) hubs, while at the same time serving as hubs for regional airports.

According to Button (2010), depending on the kind of routes being offered, airports can be classified as:

- first level airports: intercontinental and international links covering distances greater than 3,000 km; airports supporting this kind of routes are equipped to work with high traffic levels and are generally hubs for many companies;
- second level airports: international links covering distances less than 2,000–3,000 km;
- third level airports: national and international links covering distances in the range of 500–700 km.

According to the Federal Aviation Authority (FAA), airports in the USA are categorised as:[8]

- nonprimary commercial service airports are airports that have at least 2,500 and no more than 10,000 passenger boardings each year, and

[8] https://www.faa.gov/airports/planning_capacity/categories/

- primary airports commercial service airports are airports that have more than 10,000 passenger boardings each year.

Furthermore, the FAA classifies airports in the USA in four subcategories, based on the percentage of the annual US commercial boardings:

- Nonhub primary – airports handling over 10,000 but less than 0.05% of the national annual passenger boardings,
- Small hub primary – airports with 0.05 to 0.25% of the national annual passenger boardings,
- Medium hub primary – airports handling 0.25 to 1% of the national annual passenger boardings,
- Large hub primary – airports handling over 1% of the national annual passenger boardings.

Airports can also be classified according to the number of gates. Adikariwattage et al. (2012) used cluster analysis to classify airports in the US according to the number of gates and proposed four categories: small, medium, large, and very large (Table 3.5). In addition to the number of gates, the authors also proposed the following variables: passenger characteristics in terms of international/domestic passenger volumes and transfer passenger volume.

Airport classification can also be carried out based on the number of operations (number of take-offs and number of landings). According to ICAO (Annex 14), airports can be classified depending on the airport traffic density. Taking into account the number of operations (take-offs and landings) in the mean busy hour (the arithmetic mean over the year of the number of operations/movements in the daily busiest hour), the airport can have the following traffic density:[9]

- light, if the number of operations during the mean busy hour is less than 15 for each runway or less than 20 for all runways;

Table 3.5. USA airport classification according to the number of gates.

Category	Number of gates (g)	Number of US airports
Small	$1 \leq g \leq 15$	133
Medium	$16 \leq g \leq 45$	43
Large	$46 \leq g \leq 90$	14
Very large	$91 \leq g$	19

[9] ICAO, Annex 14, Aerodromes, Volume 1, Aerodrome Design and Operations, 8th edition, July, 2018.

– medium, if the number of operations during the mean busy hour ranges from 16 to 25 for each runway, or from 20 to 35 for all runways;
– heavy, if the number of operations during the mean busy hour is greater than 26 for each runway, or greater than 35 for all runways.

3.5 Airport access, catchment area, and mode of transport to the airport

Airports are usually located near cities, but at an adequate distance, to ensure unimpeded and safe traffic, bearing in mind the noise produced by aircraft during take-off and landing, obstacles (hills, mountains, tall structures), as well as the existence of access roads that connect the airport to the catchment area (Figure 3.5). The traffic between the airport and the catchment area includes not only departing and arriving passengers, but also airport staff, persons collecting and dropping off passengers, visiting the airport, delivering or picking up cargo, or going to the airport on business.

Airports can be linked to the catchment area using different means of transport, such as:

- private car,
- bus,
- rail,
- rapid transit,
- taxi, or
- tourist agency/hotel van.

Rapid transit and/or rail between the airport and the catchment area exists only if the traffic is such that it economically justifies these means

Figure 3.5. Airports in Milan and Paris.

of transport. For example, it is common for large and mid-size airports to be connected to their catchment areas by rail and/or rapid transit lines (these are airports that have an average of between 40,000 and 300,000 passengers per day). Sometimes there is a rail link if the rail station is near the airport or is part of a link to a longer line or network of lines. A bus line can transport 60 passengers per vehicle, with 3 departures per hour in both directions, from 4 a.m. to 12 a.m., which means that it can transport 7,200 passengers in both directions. Rail or rapid transit can transport between 60,000 and 400,000 passengers per day.

For example, 322,000 people travel to/from Mexico City International Airport every day.[10] The Mexico City airport is connected to a catchment area through different means of public transport: rapid transit, BRT (bus rapid transit system), which has seven lines, and bus. More than 80% of the airport staff commute using public transport, while 16% use private cars or a taxi. Of those that use public transport, half take rapid transit, 30% use the BRT or bus, 12% commute by car or motorcycle, and the rest take the taxi or walk. The distribution of trips by passengers and people accompanying them is completely different. Unlike airport staff, passengers, meeters, and greeters predominantly use private cars or taxis (around 70%), with only 20% using public transport, and the rest using long-distance buses (OECD/ITF 2018).

In recent years many airports have been promoting the use of public transport for accessing the airport, to reduce gas emissions. One of the leading airports is Oslo Airport (Norway), which has a strategy to become a green airport. This airport has managed to achieve the highest public transport utilization rate of passengers to/from the airport in Europe. In 2019 as much as 70% of the passengers used public transport (Avinor 2020).

Every day 83,000 passengers travel from Copenhagen Airport, two thirds of which are leisure passengers, and one-third are business passengers (2018). Most passengers reach the airport by public transport (61%), while 39% use private cars or taxis. Passengers predominantly use the rail (33%), one quarter uses rapid transit, and only 3% take the bus. Copenhagen Airport is one of the global leaders, with nearly 60% of the passengers using rail or rapid transit to access the airport (Copenhagen Airports 2019).

Public transport use also depends on the level of development of the transport system and public transport. The modal split by region for modes of transport to/from the airport shows that the use of private cars and taxis is the highest in the Middle East (85%) and North America (78%), as opposed to Europe (61%). The European Union has a total of 400 airports

[10] https://www.itf-oecd.org/sites/default/files/docs/surface-access-airports.pdf

with scheduled flights, of which 250 have more than a million passengers per year. Only 68 airports are directly connected to the catchment area by rail (or rapid transit) (Interreg Central Europe, LAirA project 2018).

According to the ACI, rail/rapid transit is used the most in Europe and Asia, while it is used very little in other parts of the world. In the past decade, certain airports in Europe have been developing rail infrastructure and becoming better connected to the catchment area (Barcelona in Spain, Dublin in Ireland, Lisbon in Portugal, and Prague in the Czech Republic). The focus of future development should be on airport rail links, with the aim of environmental protection. As mentioned previously, all larger airports are connected to their catchment area by rail and/or rapid transit lines, because there is a sufficient number of passengers daily, therefore this mode of public transport is economic and sustainable. At airports with more than 15 million passengers per year, the proportion of the passengers that use the rail/rapid transit to get to the airport is slightly over 30%. At airports with between 2 million and 15 million passengers per year, this proportion is around 20%.

At airports in North America, especially in the United States of America, passengers most often use private cars, rental cars, or taxis for the trip to/from the airport. Unlike in Europe, a small number of airports are linked to their catchment area by rail or rapid transit. More than ten years ago, the highest percentage of this form of travel was at the Washington D.C. Ronald Reagan National Airport (13%), Atlanta (10%), Oakland (9%), New York (JFK) (8%), San Francisco (SFO) (7%), Boston (6%), Portland (6%), Chicago Midway (5.5%), Chicago O'Hare (5%), and Newark (5%). (TRB–ACRP, 2008) Construction of infrastructure connecting Dulles International Airport with the Washington D.C. rapid transit network is expected to be completed in 2021, and this airport will be connected directly through the Silver Line (rapid transit line).

At airports with less than 2 million passengers per year, a very small portion of passengers use rail/rapid transit. The reason is that smaller airports do not have a rail/rapid transit connection to the catchment area because there are insufficient passengers. Smaller airports have this mode of transport (access) only in cases where the rail/rapid transit link to the airport is part of a long line, i.e., network of lines. Unlike large airports, small ones are mainly connected to their catchment areas by public transport lines served by local buses. Therefore, passengers reach small airports by local bus, taxi, or private car.

Airports may be located at various distances from the centre of the city. Table 3.6 depicts the distance and modes of transport to/from certain airports. It is important to note that it is travel time and not the distance that is taken into consideration because equal distance does not mean the same travel time (depending on infrastructure: highways, high-speed

Table 3.6. Distances between airports and city centres of Milan and Paris.

Airport	IATA code	Distance between airport and city centre (km)	Mode access
Milan Malpensa	MPX	54	Express train, bus, and taxi
Milan Linate	LIN	7	Bus and taxi
Milan Bergamo	BGY	53	Bus, taxi, train – Bergamo public transport
Paris Charles de Gaulle	CDG	35	Bus, train, taxi
Paris Orly	ORY	17	Bus, train, taxi
Paris Beauvais-Tillé Airport	BVA	87	Bus, train

train) as well as the mode of transport (fast train, regular train, rapid transit, bus, car). Airport accessibility is very important to passengers. The existence of different means of transport allows passengers to adequately choose how they will reach the airport.

Passengers choose differently how they will reach a given airport. The transport mode choice is carried out based on the price of the transport, travel time to the airport, travel comfort, proximity of the bus or rail station. In addition to the travel time by the given mode of transport (private car, taxi, rail, rapid transit, bus, van), the total travel time from the airport to the end destination (home, hotel, etc.) or from the trip origin to the airport also includes walking time (from the station to the terminal, from the terminal to the station, from the station in the city to the hotel or home, and vice versa), and the waiting time at the station/terminal. The travel time by private car should also include the time necessary to park and the walking time from the parking to the terminal. Figure 3.6 illustrates the overall travel time and time by segment from the airport to the hotel (hypothetical example) using different modes of transport (bus, train, taxi).

Table 3.7 gives examples of travel times and transport costs for two airports in the UK (London Heathrow and London Gatwick). As a rule, shorter travel time and greater comfort also mean more expensive transport. Longer travel time, more walking, which is usually the case with the bus, especially public transport, means a lower ticket price.

The existence of an express train linking the airport and the city centre, as a direct connection, may compete with the regular rail and/or rapid transit. Such cases exist in the case of London Heathrow, Oslo, and Vienna. Vienna has an express connection, the CAT (City Airport Train), which connects the airport with the centre of Vienna, reaching its destination in

Bus – total travel time **95 minutes**

Walking to the bus station and purchasing the ticket	Waiting for the bus departure	From the airport to the station	From the station to the hotel
15 minutes	15 minutes	50 minutes	15 minutes

Train – total travel time **70 minutes**

Walking to the rail station and purchasing the ticket	Waiting for the train departure	From the airport to the station	From the station to the hotel
15 minutes	10 minutes	30 minutes	15 minutes

Taxi – total travel time **50 minutes**

Walking to the taxi station	From the airport to the station
10 minutes	40 minutes

Figure 3.6. Illustration of total travel time from the airport to the city or hotel (hypothetical example).
Source: Kalić et al. 2020.

15 minutes. Furthermore, it is possible to check in and pick up luggage at the CAT terminal in the centre of Vienna, which makes the travel more comfortable and pleasant.

The catchment area represents the area that gravitates toward a certain airport, i.e., within which trips to/from the airport are generated. An airport's catchment area is defined based on travel time, i.e., the time necessary for the passenger to reach the airport. It is common for the catchment area to be defined as being the area within which the travel time to the airport is up to 2 hours (120 minutes). Figure 3.7 depicts the catchment area of the INI regional airport for the two-hour travel time. Greater travel time can be used in some cases, usually in connection with airports served by low-cost airlines. Low-cost airline passengers are prepared to travel even three hours to the airport, for their total travel costs (cost of air fare and costs of travel to and from airports) will be lower.

Table 3.7. Travel times and transport costs for two airports in the UK.[11]

Distance between airport and city centre	Mode of transport	Travel time and ticket price			
		Express bus	Train	Metro	Taxi
London Heathrow (LHR), 26 km	- car (private or taxi) - express train - train - metro - bus	40–50 min/ From 7 €	15 min (express)/25 € 30 min (train)/11.5 €	45–60 min/7 €	65 min/ 50–80 €
London Gatwick (LGW), 45 km	- car (private or taxi) - express train - train - bus (express and low-cost)	90–120 min/from 9 €	30 min (express)/20 € 35–45 min (train)/18.8 €	-	80–100 min/ 80–130 €

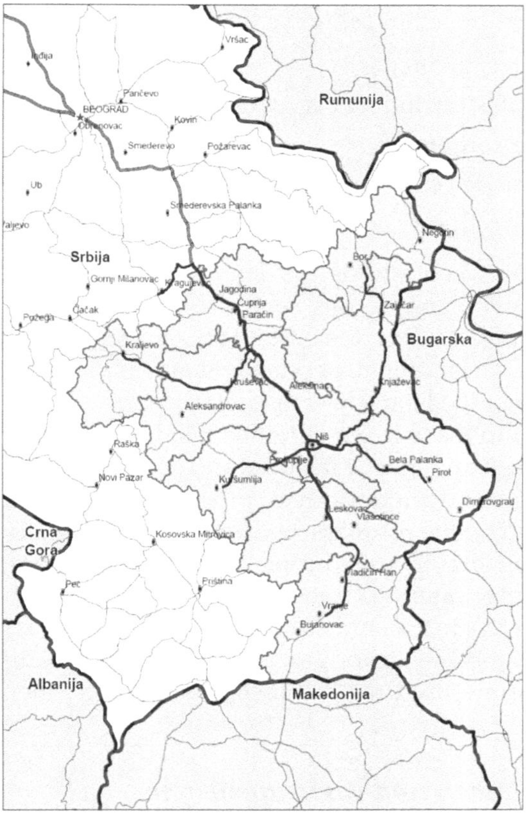

Figure 3.7. The catchment area of Airport Niš in Serbia.
Source: Kalić et al. 2020.

[11] Based on data from websites in 2019.

As mentioned above, there are areas with multiple airports and in such cases, there is great competition between the airports. As the catchment areas overlap, passengers have the option of choosing the airport from which they will fly out. Most commonly the choice is made based on the travel time to the airport, travel costs, the airport offers of destinations, and flights schedules (flight frequency, days, and times of departure).

3.6 Airport complex and operations

3.6.1 Airport complex

The airport complex encompasses the following:

1. runway and taxiway system,
2. apron(s),
3. hangars,
4. passenger terminal(s),
5. cargo complex (including cargo terminals and warehouses),
6. parking car complex,
7. transport mode access infrastructure, and
8. stations and stops (train, bus, taxi, etc.)

Every airport consists of the landside and the airside area. The runways, taxiways, and aprons comprise the airside area, intended for the movement of aircraft on the ground and aircraft ground services at the parking stands. The terminal, parking lots, access roads, and surface mode stations represent the landside area of the airport.

In addition to the above-mentioned fixed infrastructure, each airport has different fixed and mobile facilities and equipment, which are used for passenger, cargo, and aircraft service processes. This includes the lighting system near and on the runways, taxiways, and aprons, as well as various vehicles serving the aircraft during the turnaround time between two flights (refuelling, catering, transport of passengers and their baggage, power supply, etc.). In the landside area, passengers use automated people movers, moving stairs, airbridges, etc. In the terminal different facilities and equipment are used for processing passengers, their baggage, and cargo.

3.6.2 The runway and taxiway system

The runway is a defined rectangular area on a land aerodrome prepared for the landing and take-off of aircraft. Additionally, there are other areas around the runway, such as stopways, clearways, runway and safety areas, and runway strips. It is very important to provide a stopway and/or a clearway as an alternative to an increased length of the runway. The stopway is an area beyond the runway which can be used for deceleration in the event of a rejected take-off. The stopway has sufficient strength to support a decelerating aircraft in all weather conditions.[12] A clearway is an area beyond the paved runway, free of obstructions and under the control of the airport authorities. A runway and any associated stopways are included in a runway strip. A strip extends before the threshold and beyond the end of the runway or stopway. A runway end safety area is provided at each end of a runway strip. Runway and safety areas are a formal means to limit the consequences when aircraft overrun the end of a runway during a landing or a rejected take-off, or undershoot the intended landing runway. The runway strip and runway end safety area increase the safety of aircraft that undershoot, overrun, or veer off the runway, and it provides accessibility for firefighting and rescue equipment during such incidents and accidents.[13] All of the definitions mentioned above are taken from the ICAO Annex 14.

The airport can have one or more runways. There are different runway system configurations (Figure 3.8). The number of runways, as well as their configuration, depends on the number of take-offs and landings during one day, but also on meteorological conditions (wind). Generally, runways are oriented in the direction of prevailing winds. Aircraft landing is not safe in a case when the wind has a large component of the right angle to the direction of travel. Thus, crosswind can be a factor

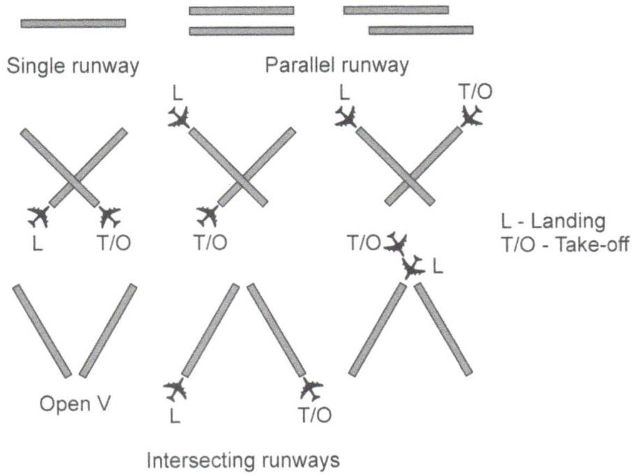

Figure 3.8. Configurations of runway systems.

[12] https://www.faa.gov/airports/airport_safety/certalerts/media/cert0003.pdf

[13] https://www.icao.int/NACC/Documents/Meetings/2014/GREPECASF1/GREPECASF1-2-3.pdf

of runway excursions. In the case of a very high probability of crosswind, two intersecting runways is a common approach in the runway system design. These runways can be used simultaneously when the wind is light. Otherwise, this configuration is used as a single-runway system (Ashford and Wright 1984). At some airports changes in the direction of the wind are pronounced during the day and for longer periods of time. Thus, multiple runways allow for an increase in capacity in terms of the number of take-offs and number of landings or allow alternate use of runways. The two runways can be parallel or intersect at a certain angle.

The six typical runway system configurations are (Janić 2007):

- Single runway;
- Two parallel runways for different modes of use (simultaneous landings and take-offs or one runway for landing and one for take-offs);
- Two converging runways, each used for both landings and take-offs;
- Two parallel plus one crossing runways, each used for both landings and take-offs;
- Two pairs of parallel runways of which the two outers are used for landing, and the two inners are used for take-offs;
- Two pairs of crossing parallel runways of which one pair is used only for landings, and the other only for take-offs.

The airports with the largest number of runways are in the USA (Chicago O'Hare (ORD) has eight runways, Dallas (DFW) has seven, Boston Logan (BOS) and Detroit Metropolitan (DTW) have six, Atlanta (ATL) has five). In Europe, Amsterdam Airport Schiphol has six runways (Figure 3.9), airports Paris Charles de Gaulle, Frankfurt, and Madrid Barajas have four runways, and airports in Barcelona and Copenhagen have three.

The runway must be long enough to allow safe operation. The following factors affect runway length: the performance characteristics of the aircraft that use the airport, the maximum take-off and landing mass of aircraft (MT/OM and MLM), the elevation of the airport, the average maximum air temperature, and the runway gradient. Usually, runway lengths vary from 2,500 to 4,000 meters at airports with scheduled commercial traffic. The runway can be even longer, which is the case of runways at the following airports: Ulyanovsk Vostochny Airport (Russia), Denver International Airport (USA), JFK International Airport in New York (USA), Madrid Barajas Airport (Spain), Toluca International Airports in Mexico, Hamad International airport in Doha, Qatar, etc. The longest runway in the world is at the Qamdo Bamda Airport which is located in

Figure 3.9. Amsterdam Airport Schiphol system runway.

eastern Tibet, China. This is the second highest airport in the world, at an elevation of 4,334 m (14,219 ft) and with a runway length of 5,500 meters.[14]

A taxiway is a path on an airport connecting a runway with the apron, hangars, terminals, and other facilities. Taxiways should be provided to permit the safe and expeditious surface movement of aircraft from the runway to the parking stand and vice versa. The design of the taxiway system is determined by the volume of air traffic, the runway configuration, and the location of the terminal building. The goal is to provide smooth and efficient aircraft flow along the taxiway (Ashford and Wright 1984).

3.6.3 Apron

The apron is a defined area on a land aerodrome, intended to accommodate aircraft for loading and unloading passengers, mail or cargo, fuelling, parking, or maintenance (Annex 14 ICAO). The apron/gate complex consists of aircraft parking stands and facilities and equipment for servicing aircraft. Its configuration depends on the configuration of the passenger terminal complex which can be linear, finger or pier, satellite, transporter (open apron), or hybrid; the movement characteristics of the aircraft to be served (e.g., turning radius); aircraft dimensions and service points; and type and sizes of ground service equipment and its usage requirements (Ashford and Wright 1984).

[14] https://www.tibetdiscovery.com/flight-to-lhasa/qamdo-bangda-airport/

There are different concepts of passenger terminal apron design (Figure 3.10). Generally, aircraft stands can be located alongside the terminal building or in an open area. Depending on terminal design (linear, finger or pier, satellite), the position of aircraft on a stand can be different (parked nose-in, parallel, or angled alongside). Remote stands (or open/remote aprons) are aircraft stands located in remote areas from the terminal.

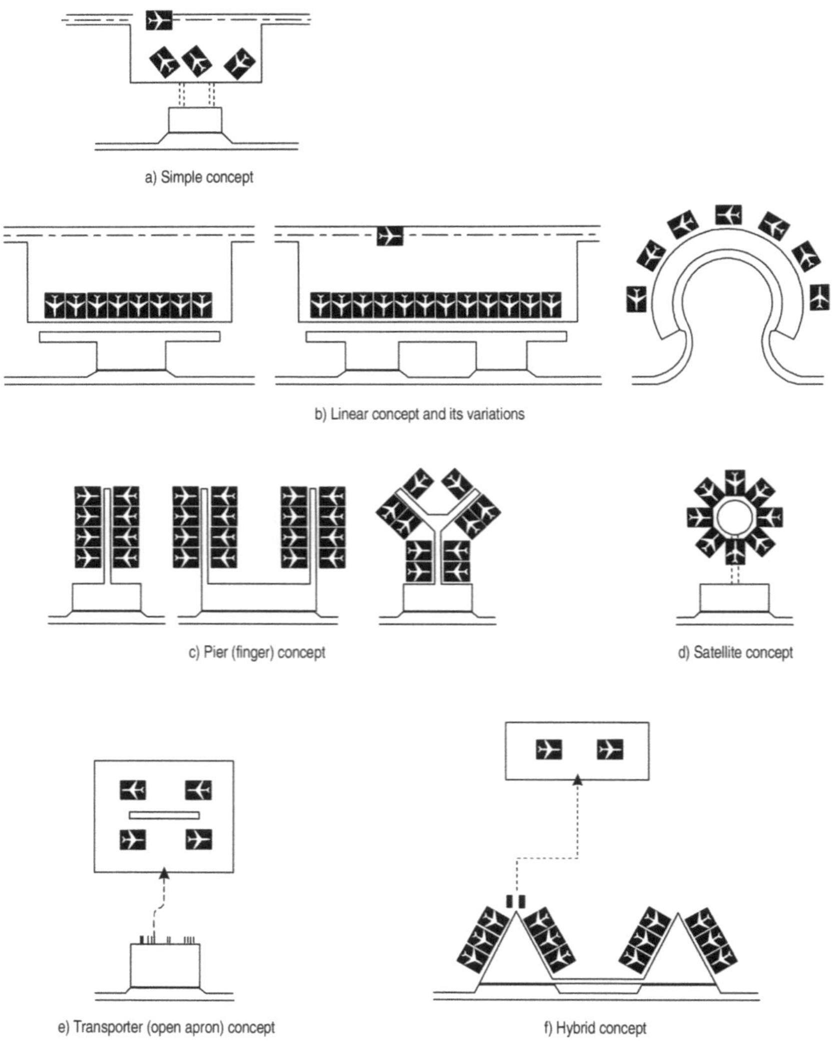

Figure 3.10. The various concepts of passenger terminal apron design.
Source: ICAO Aerodrome Design Manual - **Part 2** - Taxiways, Aprons and Holding Bays

In the case of aircraft stands located alongside the terminal building, passengers can get on the plane directly from the building, through an airbridge. If it is an open aircraft position (remote stand), special buses are provided to take the passengers to the plane. At small airports, passengers can board the plane by walking to the plane, using mobile passenger boarding stairs to enter the aircraft. In that case, the movement of passengers on the platform is strictly controlled by the airport staff.

Generally, the apron is the part of an airport used by aircraft for passenger embarcation/disembarcation and baggage/cargo loading and unloading. The ground handling process includes marshalling, choking, and connection/disconnection of ground services, baggage and freight handling, aircraft towing, refuelling, toilet and water servicing, aircraft cleaning, catering, provision of documents, high pressure air, engine and fuselage examination, and maintenance, pushback, and de-icing. The aircraft is supplied with fuel by tank vehicles or through hydrants where the fuel comes through pipes installed under the platform, which are connected to a fuel reservoir (Mirković and Tošić 2012). When the majority of the above-mentioned activities are completed, the passengers will board and luggage and cargo will be loaded.

To carry out the aircraft service process, it is necessary to have the appropriate equipment. Ground support equipment is divided into non-powered and powered. Non-powered equipment includes dollies, chocks, aircraft tripod jack, and aircraft service stairs. Powered equipment includes special vehicles and facilities such as refuellers, tugs and tractors, ground power units, buses, container loaders (lift loaders, cargo loaders or "K loaders"), transporters (cargo platforms), air start unit, non-potable water trucks, lavatory service vehicles, catering vehicles, belt loaders, passenger boarding stairs, pushback tugs and tractors, and de-icing/anti-icing vehicles.

The ground handling activities are carried out by the airport operator, airlines, or special ground handling agencies. Skilled management and staff are required due to the extreme complexity of the ground handling operations, as well as the necessity for a high level of safety and security. It is necessary to perform all ground handling activities during the expected period.

The turnaround time is the time necessary for the disembarkation of passengers from the plane and the unloading of baggage/cargo after its arrival, and its preparation for the next departure. In other words, it is the time that passes from the previous landing until the next take-off, excluding taxi-in and taxi-out. For every aircraft type, the manufacturer provides a recommended minimum duration of each of the activities within the aircraft service handling process, the sequence of activities, and a minimum turnaround time. For example, for the Embraer 195, the minimum turnaround time is 19 minutes (Mirković and Tošić 2012). The capacity of the Embraer 195 is 120 seats. In this case, the speed of deplaning is 24 passengers/minute, and the speed of boarding is 16 passengers/minute.

At congested airports, during peak hours, the ground handling operators, which are working continuously, are often not able to immediately start all the activities related to servicing the aircraft immediately after the plane lands. Therefore, it happens that the turnaround time is significantly longer, which causes delays in subsequent flights. A single delay will cause a chain reaction effect, which leads to the disruption of the airline schedule.

It is important to note that for safety reasons, the aircraft is inspected visually before each take-off. This inspection is performed by mechanics, following strictly defined aircraft maintenance procedures, looking at and checking the sensors, engines, cables, and structural elements of the aircraft.

The air side of the airport includes facilities and spaces such as:

- Hangars (buildings in which maintenance services provides repair and maintenance of aircraft as well as aircraft refurbishment).
- Airport rescue and firefighting stations, with specially designed vehicles capable of fast acceleration and high cruising speeds. The firefighters are specifically trained to extinguish fires on the plane and the most physically fit persons. According to ICAO,[15] "the operational objective of the rescue and firefighting service should be to achieve a response time of two minutes and not exceeding three minutes to the end of each runway, as well as any other part of the movement area, in optimum conditions of visibility and surface conditions." A fire on the plane must be brought under control no more than one minute after the start of extinguishing.
- Spaces with fixed equipment for defrosting and preventing icing of airplanes. In certain meteorological conditions, it is necessary to provide appropriate de-icing and anti-icing treatment of aircraft before take-off.[16] Some airports have the mobile infrastructure—special vehicles with a crane that can reach the aircraft on the parking stand and perform the de-icing and anti-icing process.

[15] ICAO document 9137 Airport Services Manual - Part1, Rescue and Firefighting, 4th edition, 2015.

[16] According to ICAO document 9640 Manual of Aircraft Ground, De-icing/Anti-icing operations, 3rd edition, 2018, de-icing is the process of removing ice, snow, slush or frost from aircraft surfaces. Anti-icing is a precautionary procedure by which clean aircraft surfaces are protected against the formation of ice and frost and the accumulation of snow and slush for a limited period.

The ATC tower, which is part of the air traffic control system, is usually located within the airside area of the airport. It is the task of the air traffic controller in the tower to manage and control the traffic at the airport, both on the ground and in the air.

3.6.4 Passenger terminal

The terminal building represents a very important part of the airport complex because it allows the passenger to check-in for flights, check-in baggage, and go through passport control, reaching the gate (the place from which the passengers board the aircraft). Also, passengers arriving at the airport by plane can "cross" the border, pick up their baggage, and go through customs control, thus completing their air travel.

Furthermore, the terminal building also includes a hall or multiple halls that have counters for checking in passengers, commercial spaces (airline offices, agencies, shops, restaurants, etc.), counters for passport control, areas for the security check of the passengers and crew members, spaces for waiting for flights, exits (gates), baggage claim areas, and spaces where passengers can be greeted.

As a rule, passengers need to be at the airport two hours before departure. The reason for arriving at the airport early is that a large number of people need to be served within a given period, where each passenger needs to be identified and issued a boarding pass, their check-in baggage received and tagged, and for them to go through passport control and security checks. The check-in closes half an hour before departure, and after that passengers cannot board the flight.

It is evident that every passenger spends time in the terminal building, which has restaurants and a large number of shops, therefore providing the terminal with features that resemble a shopping mall. For airport management, this "non-transport" segment is very important from an economic aspect, because it provides significant revenue for the airport. In addition to the passengers starting or ending their journey, the airport also includes passengers who have disembarked from the aircraft and are waiting for their connecting flights. These are transfer passengers. All the passengers at the airport should be provided a comfortable stay, a sufficiently large space with chairs, and for passengers in first class or business class, airlines provide special spaces, with a higher level of comfort. Also, there are spaces intended solely for VIP passengers.

Terminal buildings are in most cases very large, some even being so large that they resemble small cities. Some airports have multiple terminals, therefore when designing the buildings, it is necessary to take into consideration how they are connected, as well as the connections between the parts of the given building. The passenger may move through the airport on foot or using other means of transport. Larger airports are equipped with moving walkways or automated railways. Literature refers to the automatic rail system as Automated People Mover (APM). These are driverless trains that upon arrival at a stop allow the passengers to disembark on one side, and once they are empty, the doors close and the doors on the opposite side open to receive new passengers, to separate the flows of passengers embarking and disembarking from the train. On moving walkways, passengers can either stand or walk. The choice of mode of transport by passengers at the airport is affected by the walking time and standing time (on the moving walkway). With the increase in distance (distance from the gate to the terminal block), a larger number of passengers opt for the APM, with significantly fewer choosing mobile walkways and walking.

For passengers to move unimpededly and quickly through the terminal building, the passengers must be informed so that they can easily find check-in counters, terminal blocks, gates, and other areas of the airport. This is achieved through fixed signalization, placed in appropriate locations. Additionally, there are also flight information display systems that provide information related to flights (arriving and departing), locations of check-in counters, and gates. Other information may include departure or arrival times, delays, boarding, information updates.

3.7 Airport demand and capacity

Starting from the basic function of an airport, it is evident that airport service users are airlines, which use the airport infrastructure to transport passengers, cargo, and mail. Airports service aircraft on the airside and passengers on the landside. This is why airport demand is expressed as the number of passengers handled, the number of operations (sum of the numbers of take-offs and landings), and the quantity of loaded and unloaded cargo. Demand can be observed in different periods. For example, the number of passengers in a year, the distribution of passengers by day in a week, the number of passengers per hour in a day, the maximal number of passengers in an hour. Cargo and mail can also be expressed in these time intervals.

Table 3.8 shows the demand and physical capacity of London Heathrow and London Gatwick airports for the year 2019. Although airport London Gatwick has two runways, it operates as a single runway system. The second runway is not in service and is used only for emergencies. The two terminals have 245 check-in desks, 123 self-service check-in desks, and 119 self-service bag drop kiosks. The total number of passengers in 2019 was 46.6 million, of which 8.9 million were long haul passengers. The London Heathrow had more than 80 million passengers; since it is a hub airport, 94% of passengers were international and 30% were transit passengers.

Table 3.8. Demand and physical capacity of London Heathrow and London Gatwick airport.[17]

Airport/Year 2019	London Gatwick (LGW)	London Heathrow (LHR)
Number of passengers	46.6 million	80.9 million
Air transport movements (operations)	280,681	475,957
Number of terminals	2	4
Total terminal area	278,000 m2	597,463 m2 (without flight connections centre and passenger piers)
Terminals	North, South	2, 3, 4, 5 (Terminal 1 is closed in 2015)
Number of runways	1	2
Number of aircraft stands	119	212 (including 15 for cargo)

The world's busiest single-use runway is the runway at Gatwick airport, with a maximum of 55 aircraft movements per hour. In 2019, August was the busiest month with 28,000 air transport movements (take-offs and landings) at London Gatwick, and July was the busiest month at London Heathrow with 42,000 ATMs (Figure 3.11).

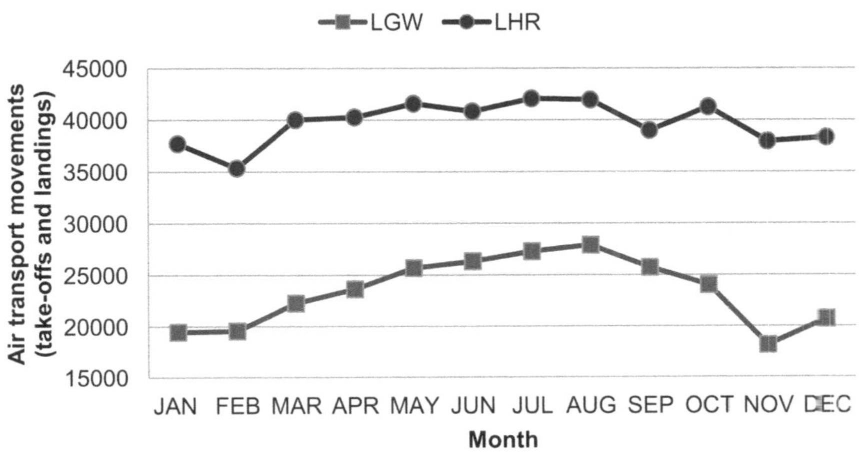

Figure 3.11. Air transport movements per month, in 2019, for London Heathrow and London Gatwick airports.[18]

[17] Based on data from: https://www.caa.co.uk/Data-and-analysis/UK-aviation-market/Airports/Datasets/UK-Airport-data/Airport-data-2019/

[18] Based on CAA data, https://www.caa.co.uk

3.7.1 Demand analysis and forecasting passenger demand

Analysis of airport traffic can be carried out at different levels of detail, depending on the purpose of the analysis and the available data. The data can be aggregated (total number of operations—number of take-offs and landings, total number of passengers, total quantity of cargo, etc.) and disaggregated (set of all individual flights with numbers of passengers, the quantity of cargo and mail onboard, and set of all individual take-offs and landings). Furthermore, it is possible to analyse the number of passengers according to the type of flights (scheduled and charter), airline (traditional and low-cost), routes (to certain destinations), etc.

Decisions pertaining to business activities that should be undertaken are passed daily at all management levels within different airport divisions, based on "yesterday's plans", to achieve "tomorrow's" expectations. "Tomorrow's" expectations are expressed through predictions, projections, or forecasts. Developing a forecast means attempting to quantify future demand (Wensveen 2015).

The development of traffic forecasts, as well as passenger demand forecasts, is important for airports. The quantified future traffic and demand are necessary for the design of airport master plans, which include infrastructure development plans (number of runways, number of parking positions and other airport elements, terminal size, number of gates, airport car parking size). Furthermore, the airport management carries out quantitative and qualitative studies with the aim of profiling passengers, for secondary airport activities related to the terminal (the offer within the terminal: shops, restaurants, and other services). The planning of airport activities—such as investment in new equipment and airport content and expansion of terminal capacities and transport infrastructure—represents a dynamic and iterative process. This means that this is a process that changes over time and demands continuous monitoring of the situation, to properly respond to the passenger demand under conditions of increasing competition.

Forecasts are categorized as short-term, mid-term, and long-term. Short-term forecasts cover a period of between a month and a year or a year-and-a-half. Predicting the daily number of passengers at the airport is also important for determining the adequate number of staff at the airport services that handle passengers and luggage. Short-terms forecasts cover airport daily operations, to properly assign employees when serving aircraft and to plan the needs for certain technical resources (gates, equipment). Consequently, short-term forecasts are also important from the aspect of planning human resources and assigning them tasks. Mid-term forecasts cover a slightly longer period, from a year to five years. These forecasts are included in tactical plans. These include investment in new equipment and new systems, and planning terminal space leasing, which is a significant source of airport revenue. Long-term forecasts cover a period from five to ten years, and even up to 15 years or more. They are used for strategic planning, which entails planning infrastructure expansion, such as the construction of new runways or new terminals, and long-term financial arrangements (cooperation with airlines and other airport users). Considering the very high investments and long periods that they cover, long-term forecasts are part of the airport's master plan.

Forecasts can have different levels of complexity. Looking at models, from the simplest to the most complex, the forecast development costs are directly proportional to their accuracy. It is believed that sophisticated econometric models are the most highly developed in a logical sense. Model sophistication implies the level of complexity and detail of the elaboration of the model, which includes a higher level of knowledge about the process that is being modelled.

The forecasting methodology is based on: relevance and availability of historical data, the desired level of accuracy, the time period for which the forecast is being created, costs and benefits of the forecast for the end user, and the time and availability of resources necessary for producing the forecast (Radnoti 2002). Creating a forecast requires knowledge and skills in the following areas: air transport sector, economics, statistics, computer programming, and communication (Vasigh et al. 2018).

To produce a forecast, it is necessary to collect the required data. Data can be obtained in different ways: some data is available in publications or online (free of charge), and some data can be ordered in the existing or a defined form. In most cases, the price of data is very high. On the other hand, some data can be obtained through one's own effort, for example by recording traffic or by passenger survey (to obtain answers from each individual passenger).

Speaking of the forecast period, which is in line with the purpose of the forecast, we use the annual number of passengers/operations, the monthly number of passengers/operations (peak month), daily number of passengers/operations (peak day), and finally peak hour. The peak period is the period with the highest number of passengers/operations. Figure 3.12 shows the distribution of passengers, arriving and departing, by the hour. There is a notable unevenness during the day, which indicates that there are periods during the day when the load is greater, and periods when the load is lesser. Also, the number of arriving and departing passengers is not evenly spread out throughout the day.

If a forecast is being developed for runways capacity, then it is operations (take-offs and landings) on the peak day and at the peak hour are taken into account. If the terminal building dimensions are being defined, then the forecasts for the peak day and peak hour are also required. The dimensions of the airport must be such that it can fully

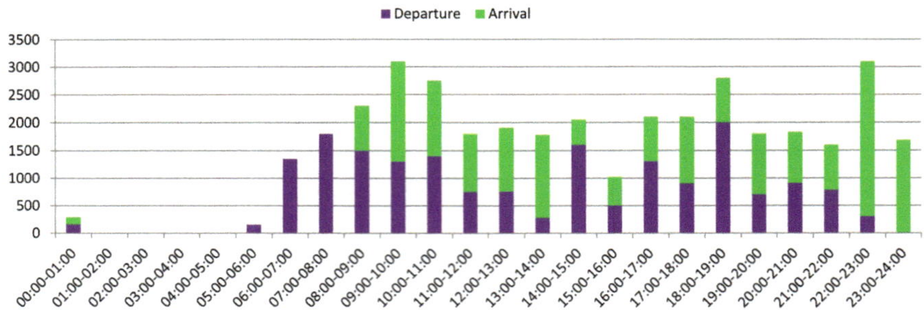

Figure 3.12. The distribution of passengers, arriving and departing, by hour.

handle its users: aircraft and passengers. This means that the flows of both the passenger arriving to and departing from the airport during the planned period should pass through the airport from the check-in counter, through passport control and security screening, to the gate and board the aircraft, and in the opposite direction, to the exit from the airport, with an acceptable queue waiting time. Passenger number forecasts for this purpose are performed also for smaller time intervals (15 minutes), and the application of certain methods can simulate the flow of passengers through the terminal.

Hypothetical example: Taking into consideration three airports—hub, non-hub, and tourist. Figures 3.13, 3.14, 3.15, and Table 3.9 show the distribution of served passengers, by month, for a hub, non-hub, and tourist airports (hypothetical examples), and their monthly stake in the total number of passengers served annually by the airport. It is apparent that the large hub does not have distinct seasonality, i.e., there is some deviation during the summer months, but it is minor. The summer months (June, July, August, and September), when residents of the northern hemisphere go on vacation, account for 37% of the total number of passengers at the given hub airport.

In the case of the non-hub airport, seasonality is more apparent, since passengers from the catchment area have greater mobility during those months, travelling for leisure, while part of the passengers from other countries is visiting the catchment area.

The summer months account for 48% of the total annual passengers served. The third example is a tourist airport, where, as expected, seasonality is prominent. As much as 76% of the total annual passengers pass through the airport during the summer months, between June and late September.

Time series forecasting. Projection represents the extrapolation of a historical trend, assuming that the factors that affect passenger demand have not changed. The number of passengers in the future is predicted based on the data on the number of passengers from the past. This functional connection

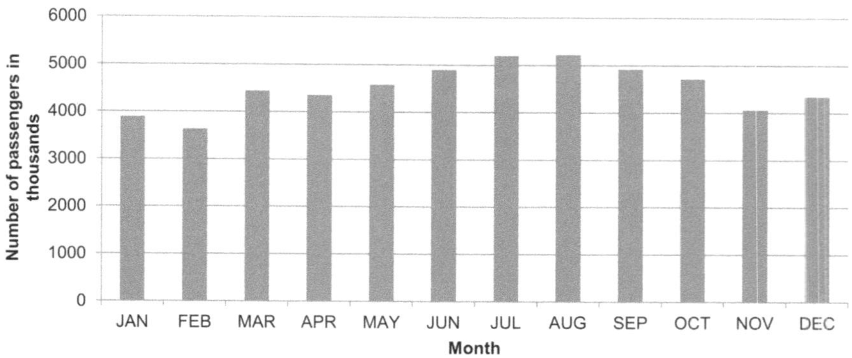

Figure 3.13. Hub airport: Distribution of the number of passengers during the year, in thousands (hypothetical example).

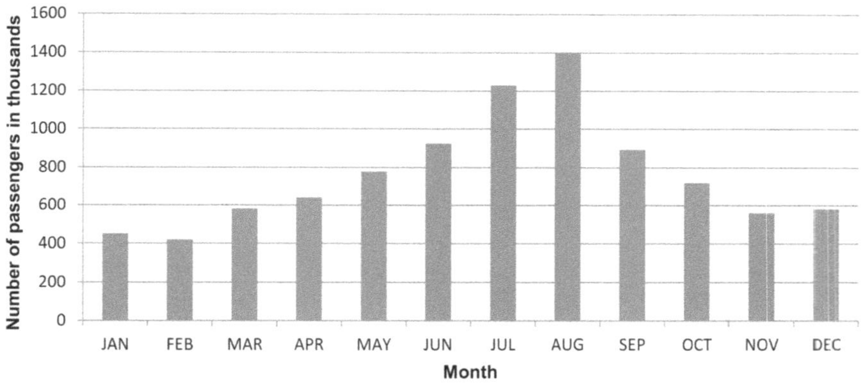

Figure 3.14. Non-hub airport: Distribution of the number of passengers during the year, in thousands (hypothetical example).

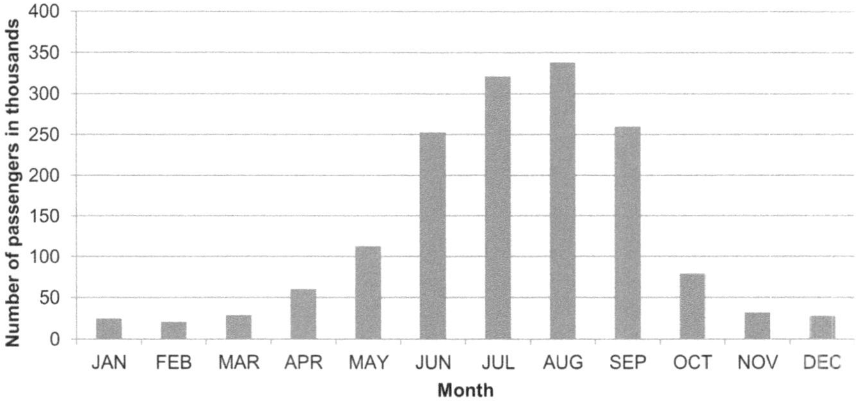

Figure 3.15. Seasonal airport (tourism destination): Distribution of the number of passengers during the year, in thousands (hypothetical example).

Table 3.9. Monthly passenger shares in the total annual number of passengers (%), hypothetical example.

	JAN	FEB	MAR	APR	MAY	JUN	JUL	AUG	SEP	OCT	NOV	DEC
Hub airport	7%	7%	8%	8%	8%	9%	10%	10%	9%	9%	7%	8%
Non-hub airport	5%	5%	6%	7%	8%	10%	13%	15%	10%	8%	6%	6%
Seasonal airport	2%	1%	2%	4%	7%	16%	21%	22%	17%	5%	2%	2%

can be linear (Figure 3.16) or an S-curve, also known as a logistic curve (Figure 3.17). In other words, the number of passengers in time can be approximated by different functions.

One of the oldest statistical methods used for forecasting is time series analysis. Time series models indicate the functional dependency between an observed variable (e.g., the number of passengers at the airport or number of operations) and a single variable–time. This technique is often used when there is not enough time available to develop a forecast or there is insufficient data available. Forecasting using time series is based on the assumption that the air traffic growth trend will continue in the future.

Time series contains four components: trends, seasonal variations, cyclic variations, and irregular fluctuations. A trend is a long-term tendency to change with time. The trend of a variable reflects the time dependency, excluding cyclic, seasonal, and irregular disruptions. Cyclic variations are variations of the observed variable over the course of a business cycle. The business cycle is a "wave" function that reflects fluctuations in economic activity. The length of a business cycle ranges from one to ten years, and it represents the period between two peaks. The business cycle has a significant impact on all segments in the air transport industry. For example, the number of business and non-business air travellers are affected by economic rises and falls, i.e., it is sensitive to changes in personal and national income. Seasonal variations are forecasted variables that depend on the time of year. They are a function of weather (vacation during the summer), as well as holidays (Christmas, Easter, Chinese New Year) during the year. Irregular variations are variations caused by random and unpredictable events (e.g., air traffic controller strikes, epidemics, natural disasters, terrorist attacks, wars, etc.) that cannot be forecasted.

Figure 3.18 depicts the number of passengers at the Nikola Tesla Airport in Belgrade, Serbia. It is evident that between 1990 and 2000 there were several irregular variations that were caused by the breakup of Yugoslavia in 1990, and two periods of UN sanctions, between 1992 and 1996, when

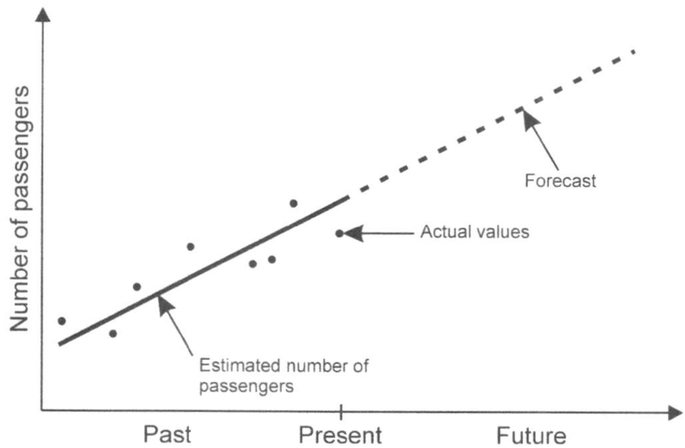

Figure 3.16. A linear connection that describes the number of passengers in time.

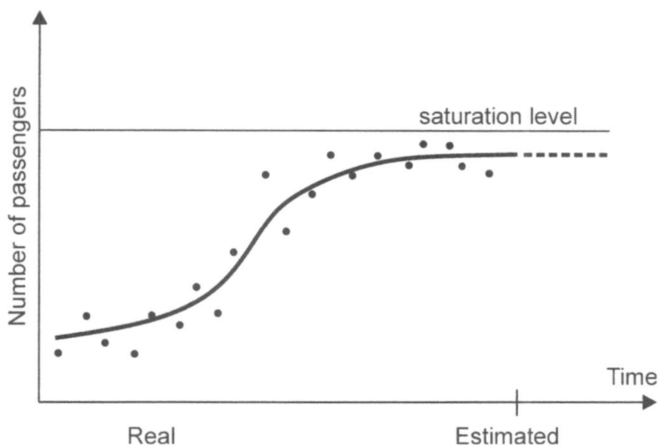

Figure 3.17. A logistic curve that describes the number of passengers in time.

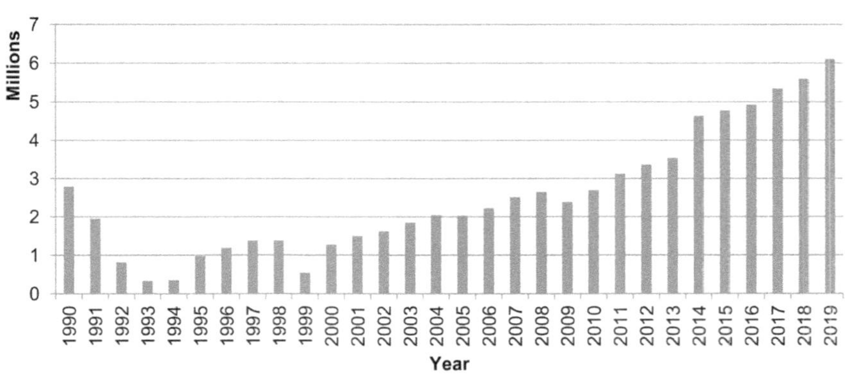

Figure 3.18. The number of passengers at BEG airport.

all flights to the European Union were banned. Furthermore, one of the greatest hyperinflations in history was recorded from 1992 to 1994. In 1999 NATO intervened in Kosovo, bombing FR Yugoslavia (Serbia and Montenegro) for three months. The 1990s were full of irregular situations, therefore it is clear that during that period the time series analysis and simple prediction were not applicable. One possibility for overcoming this issue was to eliminate the irregular years from the analysis (in this case: 1993, 1994, and 1999).

Causal models. Forecasting can be defined as the application of procedures by experts, based on qualitative and quantitative information, using relevant statistics and economic theory. Forecasts are produced based on the relationship between the volume of passenger demand and factors that affect the creation of passenger demand.

The fundamental principle of causal models is the link and impact of one or more (socioeconomic) factors on demand for air transport of passengers or cargo. For this reason, the factors (independent variables), whose impact must be considered to develop the forecast of the dependent variable (e.g., number of passengers), must be determined and selected at the very beginning. Next, it is necessary to determine the functional dependency between the dependent and independent variables, i.e., define the model that will be used. The third step is the calibration of the models (parameter adjustment) and testing the defined mathematical dependence between the dependent and independent variables. If the results of testing show a significant connection between the variables (high degree of correlation), then the last step is carried out—the final forecast. This means that it is necessary to first forecast the values of the independent variables or use existing forecasts, and then calculate the value of the dependent variable using the defined mathematical dependence.

At the global level, demand for air transport is driven by economic factors. The factors that affect demand are national economic development, international trade, and growing industrial globalisation. Furthermore, sociological and political factors, such as growing social interconnection, desire to travel, and the possibility of travel, in the sense of financial and political frameworks (disposable personal income, political restrictions, the existence of travel visas, etc.). One should also not disregard the competitiveness of other modes of transport (train, car, bus, etc.) which is reflected in price, travel time, and comfort. Demand is also affected by the offer of air transport, in the sense of destinations, price, frequency, as well as the offer of attractive trips and hotels.

$$\text{Demand}_{\text{AIRPORT}} = f\,(\text{POP, GDP, TRADE, FP, FDI, CA}_{\text{SUPPLY}}, \text{A/L}_{\text{SUPPLY}} \ldots)$$

where POP is the population, GPD is the gross domestic product, TRADE is international trade, FP is fuel price, FDI is foreign direct investment,

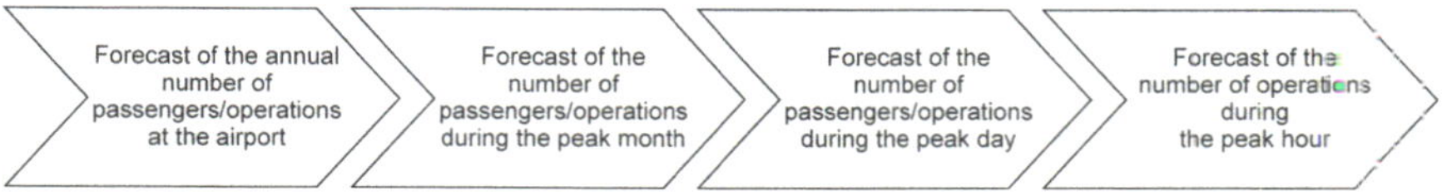

Figure 3.19. Sequential approach in producing a forecast for a single airport.

CA_{SUPPLY} is the catchment area supply, A/L_{SUPPLY} is the airline supply at the given airport.

When forecasting the number of passengers or traffic at a given airport, it is possible to apply the sequential method approach, which involves first forecasting the number of passengers/operations at the annual level (Figure 3.19). In the second step, the forecast for the peak months is produced based on the forecasted annual number of passengers and the average percentage stake of peak months during the given time period. In the third step, values for the number of passengers/operations are obtained for the peak day, and finally, the forecast for the peak hour is produced.

Finally, it should be pointed out that three things are crucial when forecasting the number of passengers and operations at an airport. Firstly, there is no single approach to airport demand forecasting; every airport has its specificities that require detailed review and examination. Secondly, the selection of the variables that affect airport demand is the most important part of the demand modelling process. Thirdly, the quality of data—meaning source reliability, accuracy and verification before it is applied—directly correlates to the forecast results.

3.7.2 Airport capacity

The term *capacity* refers to the throughput of a given service facility. In other words, airport capacity represents the maximum number of given entities that can be accommodated or served during a given period, under the given conditions (Janić 2021). Airport capacity is considered separately for the airport landside area and the airport airside area. It is important to emphasize that bottlenecks and delays can result from inadequacies in any component of the airport system—airside or landside, Ashford and Wright 1984. The capacity of the airport landside area encompasses the passenger terminal complex capacity, including the capacity of the area between the terminals and the ground access systems (rail, bus, metro, parking lots, etc.). The entities are arriving, departing, and transfer/transit passengers, their baggage, and freight/cargo. The capacity of the airport airside area includes the capacities of the runways, taxiways, and apron/gate complex that serve the entities, i.e., aircraft.

Landside area. There are two types of capacities: static ultimate and dynamic ultimate. The static ultimate capacity of the airport passenger terminal and its components can be expressed by the maximum number of passengers that can be simultaneously accommodated in a terminal and all its different parts while providing the minimum space standards assigned to each of them. The number of passengers waiting in the central terminal hall, the security checking area, etc., at a given moment in time, is estimated as $C_{st} = S/S_0$, where S is the size of the given waiting area and S_0 is the standard unit of space. Table 3.10 shows the time and space standards of the airport passenger terminals for economy class passengers, as defined by IATA.

Table 3.10. Example of the time and space standards for airport passenger terminals, for economy class passengers by IATA.[19]

Server/area	Time standard (min/pax)	Space standard (m^2/pax)
DEPARTURES		
CHECK-IN (including waiting time)		
Boarding pass and tagging	1–2	1.3–1.8
Bag drop; Desk/Station	1–5	
Check-in desk	10–20	
SECURITY CONTROL SYSTEM (including waiting time)		
Security screening	5–10	1.0–1.2
PASSPORT CONTROL (including waiting time)		
Automatic border control	1–5	1.0–1.2
Staffed emigration desk	5–10	
BOARDING GATES (50%– 70% passengers to be seated)		
Seated space	-	1.8–2.2
Standing space	-	1.2–1.5
ARRIVALS		
CUSTOM CONTROL		
Custom inspection	1–5	1.3–1.8
BAGGAGE CLAIM		
First passenger to first bag	0–25	1.5–1.7
Last bag from the first delivery	0–15	
PASSPORT CONTROL (including waiting time)		
Staffed immigration desk	5–10	1.0–1.2
Automatic border control	1–5	

[19] www.iata.org/los

The average static ultimate capacity can be estimated by assuming that the passengers spend about one hour at the terminal and that the airport operates 18h/day. The average static ultimate capacity is calculated as the ratio of the terminal area (m^2) and the average daily number of passengers (pax) per hour (h). In that case, according to Janić (2021), and data from 2018, the average static ultimate capacity for London Heathrow airport (terminal area 632,000 m^2, 80.126 million total passengers per year) is 51.8 m^2/(pax/h), for Beijing Capital International (terminal area 1,384,000 m^2, 100.983 million total passengers per year) it is 90.1 m^2/(pax/h), and for Dubai International (terminal area 1,972,000 m^2, 89.188 million total passengers per year) it is 145.3 m^2/(pax/h).

The dynamic ultimate capacity can be expressed as the maximum number of passengers that can be served during a given time period under constant demand for service. It depends on the technical and technological characteristics of the facilities. The value of this capacity is based on the average service time per customer. The dynamic ultimate capacity can be estimated as $C_{dt} = n \cdot \mu$, where n is the number of service units operating within a common service facility and μ is the average service rate of a single service unit, which is standardized. Table 3.11 shows processing times and dynamic ultimate capacities in different parts of the passenger terminal.

The scale or size of the airport is used to analyse the measure of passenger volume during the given period (Senguttuvan 2006). Airport scale is determined as the capacity to accommodate/service the physical output (number of passengers) in the airport terminal. Table 3.12 represents the average number of passengers per hour, calculated based on the number of passengers per year and assuming that the airport works 18 hours per day.

Table 3.11. Example of the processing times and processing rates in passenger terminals.

Departing passengers	**Processing time (min/pax)**	**"Dynamic ultimate" capacity (pax/h)**
Check-in	2.0–4.0	15–30
Security	0.5–1.0	60–120
Passport control	0.5	120
Boarding gate	0.5	120
Arriving passenger	**Processing time (min/pax)**	**"Dynamic ultimate" capacity (pax/h)**
Passport control	0.5–1.0	60–120
Baggage claim	2.7–5.5	11–22
Custom control	0.5–1.0	60–120

Source: Janić 2021

Table 3.12. Average number of passengers per hour.

Airport scale	Annual number of passengers in million	Average number of passengers per hour
Small	1–9	150–1400
Medium	10–30	1500–4000
Large	31–50	4700–7600
Very large	51–110	7800–17000

Airside area. The capacity of the airside area is related to components such as terminal airspace and the airport zone, the taxiway system, and the taxi lanes within the apron/gate complex. The maximum number of incoming and outgoing aircraft in a given time period (one hour) is limited by the *runway system ultimate capacity*, the taxiway system, and available aircraft gates/parking stand, under conditions of the constant demand for service. The ultimate airport airside capacity depends on many factors, which can divide into two groups: constant and variable over time. The first group consists of the airport layout, determined by the number, spacing, length, and orientation of runways; the number, locations, and design of the exit taxiways, and the design of ramp entrances.

The second group of factors, such as meteorological conditions (ceiling, visibility, and wind), aircraft mix, and type of operations, are variable over time. Additionally, air traffic control applies specific and different ATC/ATM separation rules between landing aircraft, which can be carried out either under instrument meteorological conditions (IMC) or under visual meteorological conditions (VMC).

Table 3.13 shows that IFR separation minima, applied under IMC, are 40% stricter than the VFR separation minima applied under VMC

Table 3.13. The FAA (ICAO) separation rules between the landing aircraft (in nautical miles, 1 NM = 1.852 km).

Visual flight rules (VFR)				
Type of aircraft	**Small**	**Large**	**B757**	**Heavy**
Small	1.9 NM	1.9 NM	1.9 NM	1.9 NM
Large	2.7 NM	1.9 NM	1.9 NM	1.9 NM
B757	3.5 NM	3.0 NM	3.0 NM	2.7 NM
Heavy	4.5 NM	3.6 NM	3.6 NM	2.7 NM
Instrument flight rules (IFR)				
Type of aircraft	**Small**	**Large**	**B757**	**Heavy**
Small	2.5 (3) NM	2.5 (3) NM	2.5 (3) NM	2.5 (3) NM
Large	4.0 NM	2.5 (3) NM	2.5 (3) NM	2.5 (3) NM

Table 3.14. Examples of ultimate capacities of runway systems.

European airports			
Airport/Code/Country	**Number of runways**	**Declared IFR capacity (atm/h)**	
Paris/CDG (France)	4	112	
London Heathrow/LHR (UK)	2	82	
Amsterdam Schiphol/AMS (Netherlands)	6	106	
Airports in U.S.A.			
Airport/Code/Country	**Number of runways**	**VFR capacity (atm/h)**	**IFR capacity (atm/h)**
Atlanta/ATL (Georgia, USA)	4	180–188	158–162
Chicago O'Hare/ORD (Illinois, USA)	7	190–200	136–144
Phoenix Sky Harbor/PHX (Arizona, USA)	3	128–150	108–118

Source: Janić 2021

(Janić 2013). The separations guarantee that there is no risk. The ultimate capacity of the runway system at selected airports in Europe and the United States is presented in Table 3.14.

According to Janić (2021), the *ultimate capacity of the taxiway system* represents the maximum number of taxing aircraft that can pass through the "bottleneck" segment, as a link and/or intersection node on the taxiway connecting the runway and apron/gate complex.

The *ultimate apron/gate complex capacity* can be static and dynamic. Static capacity is related to the number of available aircraft stands/gates for parking particular aircraft categories. The dynamic capacity refers to the maximum number of aircraft during a time period (one hour), under constant demand for service. It depends on the physical characteristics of the apron/gate complex (number of parking stands by aircraft type or by airlines), as well as taxiway system, the corresponding service time of particular aircraft categories/flights, and demand characteristics such as flight distribution (arrival and departures) in time, flight type by airlines, etc.

3.7.3 Matching airport capacity and demand

Matching the airport capacity and demand represents a crucial task for airports management, both at the short-term operational level and in long-term planning. The short-term implies time periods such as hours, days, months, or a year, and refers to airports procedures and use of equipment and resources. According to Janić (2013), the main problem of airports is finding an efficient and effective way of matching their fixed capacity to growing demand.

At the operational level, it is exceptionally important how many passengers and aircraft there are in the peak hour (the hour with the highest demand). This is why take-off and landing times at an airport are planned in advance, so that existing resources may service both passengers and aircraft within the given time. Furthermore, there is also the capacity linked to the number of parking positions where a certain number of aircraft can be parked at one time. Particularly important are the human resources involved in servicing the passengers and aircraft. One way for dealing with delays at large airports that often experience congestions is the introduction of airport slots—predefined time periods when the airline can use airport infrastructure. Airlines obtain slots at such airports before every season. The process of matching the demand and airport capacity at the operational level is based on the flight schedule (the distribution of flights in the day, at the given airport), the terminal capacity (number of check-in counters, number of passport control counters, security control capacity, the capacities of shared spaces and gates, baggage claim area), and the capacity of the airside (runway system, taxiway system capacity, number of parking positions, aircraft handling service capacity, etc.).

In the process of matching the airport capacity and demand, long-term planning is related to physical expansion of the infrastructure in the airport airside and landside area, such as new terminals, new cargo terminal complexes, new runways, additional apron/gate complexes, etc. Usually, the long-term period spans 15 to 20 years. Physical expansion of the airport infrastructure is usually carried out in a step-by-step manner, in accordance with the Airport Master Plan. Each Master Plan consists of airport traffic and passenger demand forecasts (hourly, daily, monthly, and yearly volumes). Based on forecasts and expected demand growth, the Master Plan defines the airport's future development, taking into account large capital investments into airport infrastructure.

3.8 Airport business and economics

3.8.1 Airport business

It used to be that, in a business sense, airports served solely for conducting traffic. The end product was the service that was provided to airlines and their passengers, and which entailed the use of the airport infrastructure. For passengers, the airport represented the origin of their travel, where they handed their luggage and boarded the airplane, and the destination, where they disembarked and picked up their luggage. For the airline, this was the place that enabled aircraft take-off and landing, as well as aircraft handling. Traditional airport operation was based on state ownership of the entire infrastructure and a simple relationship between the airport and the airline (Figure 3.20). Airports focused on revenues from airport charges for infrastructure used by the airlines, as well as charges for serving passengers and aircraft. At that time there were only duty-free shops, restaurants, and cafés, but significantly less than there are today.

In the 1990s airports became commercialised and the business model changed. Unlike traditional operations, airports now focused on passengers, by branching out their operations into the commercialisation of the terminal space, therefore significantly increasing non-aeronautical revenue. Airports recognised that the free time that the passengers have between going through passport control and boarding the aircraft (waiting time), represents commercial potential, so they took advantage of the capacities of the space in the terminal and made it more attractive for shopping and spending time in cafés and restaurants. In addition to the passengers, the airports also focused their business on terminal shop leaseholders and car parking providers. Airports are also linked to hotel operators, i.e., hotel chains that use hotels (realty) owned by the airport and which are very close to the terminal. Some airports have created the image and advertise themselves as "airport cities", to emphasize the broad range of services that they offer (Humphreys and Francis 2017).

Modern airport business implies a "multipoint firm" concept (Figure 3.21). According to Jarach (2001), "this kind of enterprise seems, in its best practices, capable of serving various demand clusters with an array of different value propositions: attractions for tourists and aviation

Figure 3.20. Traditional airport business.

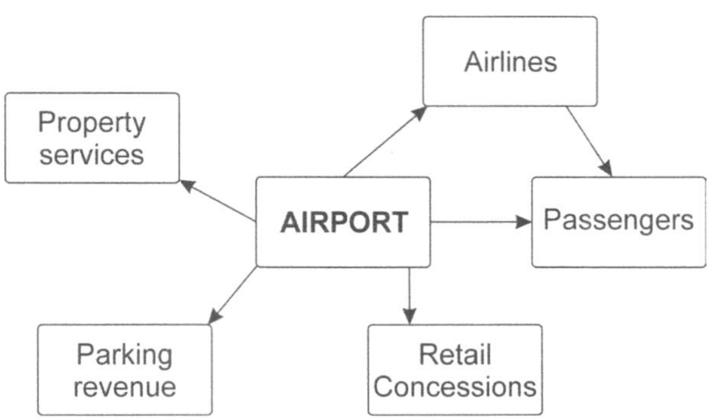

Figure 3.21. Modern airport business.

enthusiasts, congressional and logistic facilities, shopping venues and consulting services appear the most relevant ones." Thus, airport executives have to pay attention to their business performance, as well as understanding passenger experience at the airport, which has become essential for airport management (Bezerra and Gomes 2020).

With the increase in competition between airports, transfer passengers have gained a greater choice of airports. The time that transfer passengers spend at an airport waiting for their next flight represents an opportunity for those managing the airport and its content. In addition to the traditional approach of dividing passengers into business and non-business passengers (leisure, visiting friends and relatives, etc.) there is also a classification based on socio-demographic characteristics (age, gender, origin in the sense of cultural attributes), or based on behaviour and shopping decision-making style. Depending on the passenger segment at the airport, management creates an adequate offer, to maximize airport revenue.

The airport creates its business depending on core activities and customers profile (different types of airlines, business or leisure passengers, etc.). According to Jarach (2001), airports can be classified into five groups, according to the type of traffic and business model:

- primary hubs: this type of airport is located within the service area of large economic centres that generate large demand (origin-destination) and represent a hub for one or more airlines, therefore they also have large transfer traffic; at such airports flights (short- medium, and long-haul) are operated by airlines with different business models (full service carriers, low-cost carriers, regional carriers, etc.);
- secondary hubs: unlike primary hubs, this type of airport has a slightly lower traffic volume, with a more pronounced focus on short-haul flights, although medium- and long-haul flights exist; it is a hub for lower density networks, but with higher flight frequency; these airports also serve airlines with different business models;
- regional airports: this type of airport focuses on point-to-point services, and it also includes so-called city airports; such airports serve domestic and international flights and are linked to hubs;
- low-cost carrier airports: such airports are used by LCCs and services provided at them are adjusted to lower prices, to enable LCCs to offer cheaper airline tickets;
- cargo airports: these airports are completely adapted to the requirements of cargo airlines; the technical infrastructure includes all the equipment required for cargo manipulation and handling.

3.8.2 Airport ownership and management

The 1990s featured changes regarding the ownership and management of airports, with a shift in the direction from the public sector toward the private sector. Some airports were completely or partially privatised. Partial privatisation entailed minority or majority state ownership. When discussing the privatisation of airports, we differentiate between two separate processes: airport ownership and airport management. The ownership directly affects airport management.

Airports that are entirely publicly owned may be owned by the state national government (state ownership) or owned by the provincial, regional, or local government. Airports can also have mixed ownership, partially owned by the state, and partially by the provincial, regional or local government. In such cases, airports are managed and controlled by the state, through the civil aviation authority and the ministry of transport. The financing of airports, i.e., airport infrastructure, through capital investments, is limited by the state's borrowing power.

The environment, other commercial sectors, the transport policies of other states and airlines have prompted states to consider the privatisation of airports. Looking at the global growth of the number of passengers, year after year, and bearing in mind the costs of investment in the expansion of airport infrastructure, privatisation appeared to be a good solution.

Certain countries have opted for the public sector model (state ownership of airports) with the participation of the private sector (management or contracted services, alternative financing, etc.). In addition to this model, there is also the combination of state and private ownership, where the state may be the majority or minority owner. Also, an increasingly common system involves management by independent or semi-dependent bodies—airport operators, which has led to further commercialisation of airports. For example, Fraport is a German operator that manages Frankfurt Airport, and also manages other airports, for a contracted period (Indira Gandhi International Airport in India, 2006–2036; Antalya Airport in Turkey, 1999–2024; the airports in Burgas and Varna, Bulgaria, 2006–2041; 14 regional airports in Greece (Aktion, Chania, Corfu, Kavala, Kefalonia, Kos, Mykonos, Mytilene, Rhode, Samos, Santorini, Skiathos, Thessaloniki, Zakynthos) 2017–2057; Ljubljana in Slovenia, since 2014 and indefinitely, Pulkovo Airport in Saint Petersburg, Russia, 2010–2040; Xi'an Xianyang International Airport in China, retail at several US airports (Baltimore, Cleveland, Nashville, New Jersey, New York, Pittsburgh), Jorge Chávez International Airport in Lima, Peru, 2001–2041; the airports in Fortaleza (2018–2047) and Porto Alegre (2018–2042) in Brazil). The company also owns or co-owns several airports, with its stake ranging from 10% to 100% (Fraport Annual report 2020).

According to an Airport Council International (ACI) report from 2017, of the 100 airports with the highest traffic in the world, 46 have some form of private sector participation. The proportion of passenger traffic handled by airports with private sector participation, by region, is as follows: 75% in Europe, 60% in Latin America and the Caribbean, 45% in Asia and the Pacific, 13% in the Middle East, 11% in Africa, and 1% in North America. Europe has the largest propensity for private sector involvement in terms of passenger traffic handled by fully privatised airports or public-private partnerships (PPPs) as well as has the highest absolute number of privatised airports (266).

In the case of the USA, in many cases, airlines lease terminals partially or entirely. In Australia too, local airlines lease terminals, with the exclusive right to use them. That way airlines have control of the terminal and the airport's entire operation.

3.8.3 Airport economics

Airport economics refers to revenues, costs, profits, and losses. The profit (or loss) is calculated as the difference between the total revenues and total costs. The profit can be expressed in absolute terms, per year. Airport size plays an important role in determining profitability. In order for an airport to achieve economies of scale, total costs need to be spread out over the airport's outputs (i.e., passengers, cargo, and air transport movements—operations). In the airport's financial report, net profit is obtained as the difference of EBITDA (earnings before interest, taxes, depreciation, and amortization) and all expenses from revenues (depreciation, financial costs, and taxes).

Airport costs. The total airport costs are the sum of operating and capital costs. The airport cost structure consists mainly of high fixed costs of maintaining and operating the infrastructure components. In 2016, operating expenses were 64.7% of total costs, and the rest were related to capital costs (ACI Media Releases 2018).

The operating costs cover personnel, contracted services, communications, utilities, administration, maintenance, lease, rent, concessions, materials, equipment, insurance costs, and other costs. Two-thirds of the total airport costs are related to personnel costs. One quarter of the total operating costs is for contracted services.

Capital costs are related to infrastructure development. These are investments in fixed assets or improving existing assets (equipment, infrastructure, property, land, etc.). The capital costs consist of depreciation and/or amortization, interest, and other capital costs.

Airport revenue. There are two types of airport revenues: aeronautical and non-aeronautical. According to data available in the literature, on average, around 55% of total airport revenue is from aeronautical services, around 40% is from non-aeronautical services, and close to 5% is non-operating revenue (these are the average values for the 2005–2016 period). If we look at individual cases, the distribution of revenues differs greatly from airport to airport. In some cases, 75% of the revenue is from aeronautical charges (e.g., the airports in Vienna, Salzburg, and Florence), and in other cases, it is much lower than the average (e.g., Dublin Airport, around 38%) (Graham 2009).

Aeronautical revenue is generated based on conducted traffic, from servicing passengers, cargo, and luggage on one hand, and aircraft servicing, take-off, and landing on the other. Aeronautical revenue stems from the following charges: landing charges, passenger charges, cargo charges, aircraft parking charges, security charges, noise charges, charges for the use of infrastructure (jet bridge, luggage handling system, check-in counters, infrastructure for passengers with reduced mobility, de-icing facilities, hanger usage, if it is owned by the airport, etc.). According to data from 2016, globally, on average, passenger charges account for the majority of aeronautical revenues (45%), followed by landing charges (21%), terminal leases (which primarily applies to the USA, 12%), security charges (8%), etc., Figure 3.22 (ACI World Seminar 2018).

Airport charges differ from airport to airport. Large hubs, which have very high traffic volume, as a rule, have higher airport charges (e.g., London Heathrow, Amsterdam, Zurich), and vice versa—airports with lower traffic volume or airports that have higher capacity than the existing traffic, strive to attract airlines by offering lower airport charges.

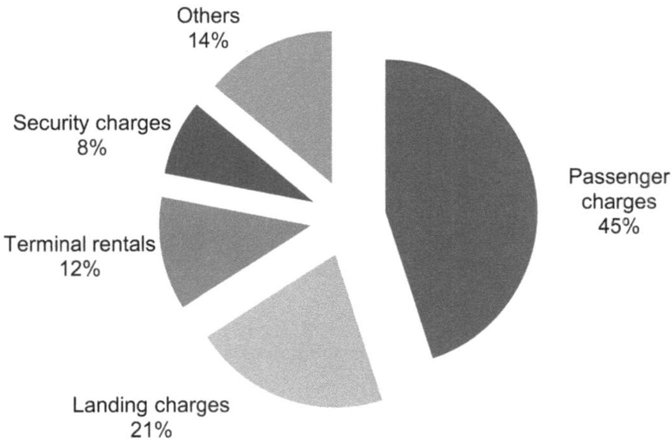

Figure 3.22. Distribution of aeronautical revenue by source (2016).

As stated in Graham (2009), there are two key factors for increasing non-aeronautical (commercial) revenue at airports: first, airports have evolved from the public sector to the commercial, and in some cases to the private sector, which has motivated airports to explore possibilities and offer commercial content; second, there is pressure from airlines on the airport regarding the aeronautical charges.

Non-aeronautical revenues include all airport revenues that are not related to traffic (aircraft operations and servicing passengers). The sources of non-aeronautical revenue are retail concessions (sales of goods or services, restaurants, cafeterias, and other similar businesses), airport car parking (concessions, airport-owned operator-managed, or airport-owned airport-managed), rent (leasing real estate or commercial space), and providing consulting services.

Figure 3.23 shows the distribution of non-aeronautical revenue according to category, globally. Around two-thirds of non-aeronautical revenue is related to retail concessions, car parking, and realty ownership (concessions, leasing, own revenue).

Figure 3.24 shows the distribution of non-aeronautical revenue according to category and region. It can be noted that in the Middle East retail concessions accounts for as much as 56% of revenue, which is far more than in any other region. The global average for this category is 29%, and it is only 8% in North America, ACI World Seminar 2018. On the other hand, parking concessions and parking owned by the airport accounts for the majority of the non-aeronautical revenue in North America. Compared

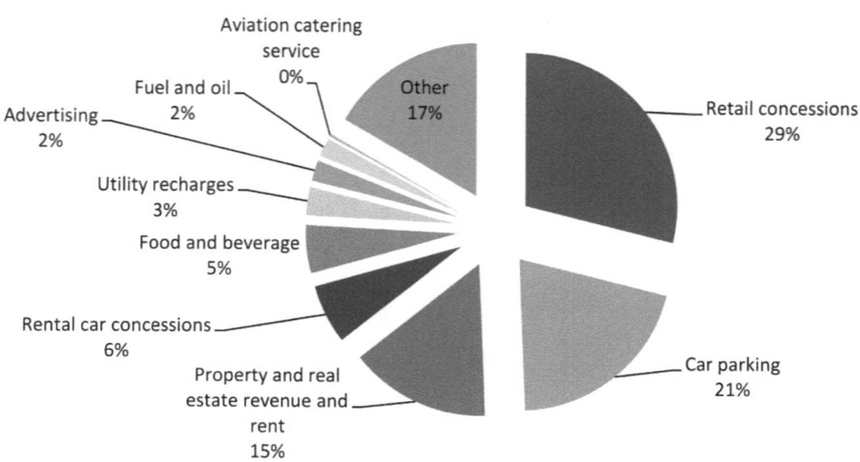

Figure 3.23. Distribution of non-aeronautical revenue by source (2016).[20]

[20] https://www.icao.int/Meetings/aviationdataseminar/Presentations/ACI%20World_Seminar_Paris.pdf

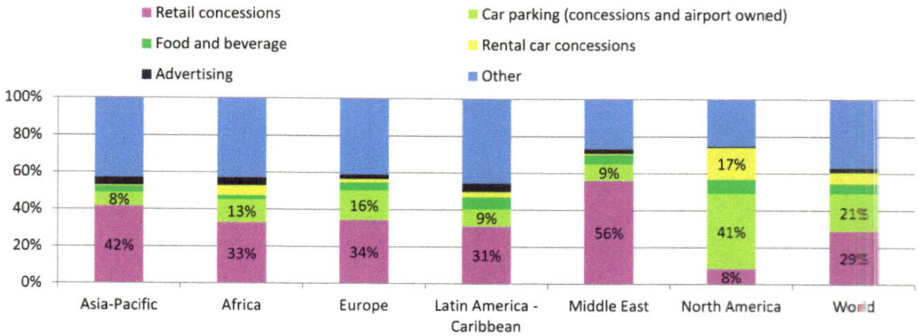

Figure 3.24. Distribution of non-aeronautical revenue by source (2016).[21]

to the rest of the world, North America has a very high portion of car rental concessions in the total non-aeronautical revenue (17%), while in other regions this stake is far less significant (1% to 3%, and 6% only in Africa). If we combine the retail concession revenue and revenue from car parking, we can conclude that in most cases they account for 50% of non-aeronautical revenue. The only regions that diverge from this value are Latin America and the Caribbean (40%) and the Middle East (65%) (ACI World Seminar 2018).

Incentive programs offered by airports. To survive in the new market conditions (market liberalisation and airport privatisation) many airports around the world offer certain discounts on airport charges that they charge airlines. In the past decade, airports around the world have developed different types of measures aimed at encouraging development and traffic growth. Today there are a series of different incentive programs for airlines, proven and improved in practice, aimed at generating additional traffic.

For example, at the Brussels Airport, the (basic) passenger charge is € 21.74, for transfer passengers it is € 11.17, which is increased by € 0.62 for bus transport on the apron if it exists.[22] According to the airport's offer, which is valid from April 2016 to March 2021, if a new long-haul flight is introduced (a new destination), a discount is provided on the passenger charges, which for passengers starting their trip at this airport amount to € 15 during the first year, € 10 during the second, and € 5 during the third year. For transfer passengers the charge is € 7.50 during the first year, € 5 during the second, and € 2.50 during the third year. Also, this airport provides incentives for airlines that have a positive passenger growth trend, and in that case, the charge is € 10 for passengers who start their

---

21 https://www.icao.int/Meetings/aviationdataseminar/Presentations/ACI%20World_Seminar_Paris.pdf

22 Charges & fees at Brussels Airport, https://media.brusselsairport.be/bruweb/default/0001/33/9a5dc00b93038c7517d870bf1a99658f9a562512.pdf

trip at the airport, and € 5 for transfer passengers. Also, if the number of cargo flights increases (the flight frequency is increased), the landing and parking charges are reduced 25% the first year, 50% the second year, and 75% the third year.[23]

References

ACI Media Releases. 2018. ACI World's Airport Economics Report illustrates strength of global industry as aviation's center of gravity continues to shift eastwards, April 10, 2018.

ACI Statistics Manual. 2011. A Practical Guide Addressing Best Practices.

ACI World seminar. 2018. Lucas, P., Short stories in Airports Economics (presentation), Paris, April 18th, 2018. https://www.icao.int/Meetings/aviationdataseminar/Presentations/ACI%20World_Seminar_Paris.pdf.

Adikariwattage, V., de Barros, A.G., Wirasinghe, S.C. and Ruwanpura, J. 2012. Airport classification criteria based on passenger characteristics and terminal size. Journal of Air Transport Management 24: 36–41.

Ashford, N. and Wright, P.H. 1984. Airport Engineering, 2nd Edition, John Wiley & Sons, Inc., New York/Chichester/Brisbane/Toronto/Singapore.

Avinor. 2020. Environmental Report 2019, Oslo Lufthavn, Gardermoen, April 2020. https://avinor.no/en/corporate/airport/oslo/community-and-environment/miljorapporter/.

Bezerra, George, C.L. and Gomes, Carlos, F. 2020. Antecedents and consequences of passenger satisfaction with the airport. Journal of Air Transport Management 83(C): 101766.

Brussels Airport, Charges & Fees at Brussels Airport, https://media.brusselsairport.be/bruweb/default/0001/33/9a5dc00b93038c7517d870bf1a99658f9a562512.pdf.

Brussels Airport, Incentive Program (applicable from 1 April 2016 until 31 March 2021).

Button, K. 2010. Chapter 1 Economic aspects of regional airport development. *In*: Postorino, M.N. (ed.). *Development of Regional Airports: Theoretical Analyses and Case Studies*. University of Reggio Calabria, Italy.

CNN Travel. 2019. This is the world's busiest airport, https://edition.cnn.com/travel/article/worlds-busiest-airports-2018/index.html.

Copenhagen Airports. 2019. Take off for a sustainable future, Group Annual Report 2018. https://www.cph.dk/495470/globalassets/8.-om-cph/6.-investor/arsrapporter/2018/cph-annual-report-2018_uk_web.pdf.

Eurostat, Press Release, 186/2019– 6 December 2019, Eurostat Press Office, https://ec.europa.eu/eurostat/documents/2995521/10265946/7-06122019-AP-EN.PDF/8f2c9d16-c1c4-0e1f-7a66-47ce411faef7.

FAA, Airport Categories, https://www.faa.gov/airports/planning_capacity/categories/.

Fraport, Annual Report. 2020. https://www.fraport.com/en/newsroom.html.

Graham, A. 2009. How important are commercial revenues to today's airports? Journal of Air Transport Management 15: 106–111.

https://aci.aero/news/2018/04/10/aci-worlds-airport-economics-report-illustrates-strength-of-global-industry-as-aviations-center-of-gravity-continues-to-shift-eastwards/#:~:text=Cost%20and%20investment&text=In%202016%2C%20operating%20expenses%20made,investments%20in%20runways%20and%20terminals.

[23] Brussels Airport, Incentive Program (applicable from 1 April 2016 until 31 March 2021).

https://www.brusselsairport.be/en/aviation-development/incentive-program#:~:text=The%20objective%20of%20the%20incentives,that%20serves%20a%20new%20destination.

Humphreys, I. and Ison, S. 2005. Changing airport employee travel behaviour: the role of airport surface access strategies, Transport Policy 12: 1–9.

Humphreys, I. and Francis, G. 2017. The airport-airline relationship. *In*: Budd, L. and Ison, S., Routledge (eds.). *Air Transport Management: An International Perspective*. Taylor & Francis Group, London and New York.

IATA, Level of Service Concept, https://www.iata.org/los/.

ICAO region, ICAO and IATA codes https://www.icao.int/wacaf/documents/apirg/sg/2012/afi_opmet_mtf4/docs/wp10_atta.pdf.

ICAO. 2015. document 9137 Airport Services Manual - Part1, Rescue and Firefighting, 4th ed.

ICAO. 2018. Document 9640 Manual of Aircraft Ground, De-icing/Anti-icing operations, 3rd ed.

ICAO. 2018. Annex 14, Aerodromes, Volume 1, Aerodrome Design and Operations, 8th ed.

ICAO. 2020. Annex 14, Doc. 9157, Aerodrome Design Manual, Part 1, Runways, 4th ed.

Interreg Central Europe. 2018. Air-Rail Link - a LAirA Project Report, EU European Regional Development Fund, https://www.interreg-central.eu/Content.Node/Action-Plan-1.pdf.

Jarach, D. 2001. The evolution of airport management practices: towards a multi-point, multi-service, marketing-driven firm. Journal of Air Transport Management 7: 119–125.

Janić, M. 2007. The Sustainability of Air Transportation: A Quantitative Analysis and Assessment, Taylor & Francis Ltd., Aldershot, England, United Kingdom.

Janić, M. 2013. Airport Analysis, Planning and Design: Demand, Capacity and Congestion. Novinka, Nova Science Publishers, Inc. New York.

Janić, M. 2021. System Analysis and Modelling in Air Transport: Demand, Capacity, Quality of Service, Economics, and Sustainability, CRS Press, Taylor & Francis Group.

Kalić, M., Dožić, S. and Babić, D. 2020. Osnovi vazdušnog saobraćaja, Univerzitet u Beogradu – Saobraćajni fakultet, Beograd, I izdanje.

Mirković, B. and Tošić, V. 2012. Vazduhoplovna pristaništa, Univerzitet u Beogradu – Saobraćajni fakultet, Beograd, I izdanje.

OECD. 2018. Surface Access to Airports: The Case of Mexico City's New International Airport. International Transport Forum, https://www.itf-oecd.org/sites/default/files/docs/surface-access-airports.pdf.

Radnoti, G. 2002. Profit Strategies for Air Transportation, Aviation Week Books, McGraw-Hill.

Senguttuvan, P.S. 2006. Economics of the Airport Capacity System in the Growing Demand of Air Traffic – A Global View, Transport Research Annual Forum, New York University, March 2006.

TRB - ACRP, National Academies of Sciences, Engineering, and Medicine, 2008. Ground Access to Major Airports by Public Transportation. Washington, DC: The National Academies Press. https://doi.org/10.17226/13918.

Vasigh, B., Fleming, K. and Tacker, T. 2018. Introduction to Air Transport Economics: From Theory to Applications, Routledge, 3rd ed.

Wensveen, J. 2015. Air Transportation: A Management Perspective. Routledge, 8th ed.

Chapter 4

Air Traffic Control System

Air traffic control (ATC) is a service provided by ground-based controllers who direct aircraft on the ground and in the air. Air traffic controllers are people who operate air traffic control systems to maintain a safe and orderly flow of air traffic and help prevent mid-air collisions, as well as to ensure safe separation of aircraft with the obstacles (e.g., terrain, buildings, etc.).

4.1 Air traffic management

Air traffic management (ATM) means the aggregation of the airborne and ground-based functions required to ensure the safe and efficient movement of aircraft during all phases of operations (EC Regulation 549/2004). It is designed to assist commercial aircraft in travelling between airports, as well as during the procedures of landing, taking off, and manoeuvring at the airport.

Main services provided to air traffic during all phases of operations (approach, aerodrome, departure, and en-route) include air traffic management, communication, navigation and surveillance, meteorological services for air navigation, search and rescue, and aeronautical information services (ICAO Annex 11 2018). To ensure safe, economical, and efficient management of air traffic, ATM provides the facilities and seamless services in collaboration with all parties and involving airborne and ground-based functions (ICAO Doc 4444 2016). Thus, to achieve the highest degree of uniformity and to enable and improve safety and efficiency, the ICAO Council adopted the International Standards and Recommended Practices where procedures for safe and efficient air navigation services are approved.

ATM as dynamic, integrated management of air traffic and airspace, comprises three main services (Figure 4.1):

- Air traffic services (ATS);
- Airspace management (ASM) and
- Air traffic flow management (ATFM).

There are three main tasks that an ATC service provider is obligated to enable and they involve the safe, orderly and expeditious flow of air traffic within airspace under the sovereignty of a state. *The safe conduct of flights* means prevention of collisions between aircraft, and as prevention of collisions between aircraft on the manoeuvring area and obstructions in that area. *Maintaining an orderly flow of air traffic* means creating such conditions for traffic operations in which aircraft delays are reduced to a minimum. *Expeditious flow of air traffic* means that the declared capacity of the air traffic control services concerned exceeds air traffic demand to be able to meet demand requests and to guarantee an orderly flow of traffic (Babić and Netjasov 2016).

Air traffic services (ATS). ATS are established to ensure safe and orderly traffic flow within airspace or at an aerodrome (ATC service), to provide necessary information to flight crews (flight information service, FIS) and, in case of an emergency, to alert the appropriate bodies (alerting service), Figure 4.1. ATS are provided by specially designated air traffic service units (ATSUs) and these units may provide more than one type of service. ATS are mostly performed by air traffic controllers. Their main functions are to prevent collisions between aircraft, to prevent collisions between aircraft on the manoeuvring area and obstructions, and issue timely clearances and instructions to maintain an orderly flow of air traffic (ICAO Doc 4444 2016). ATS are based on tactical interventions by controllers (e.g., applying appropriate separation standards) and direct the communication with flight crews usually during the entire flight (e.g., receiving crew requests for desired levels and flight paths, reducing holding times in the air and on the ground, etc.).

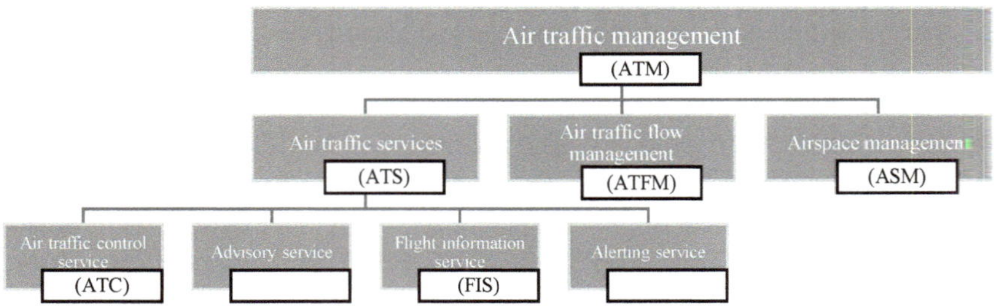

Figure 4.1. Structure of ATM and the relations between ATM, ATS, and ATC.

Air traffic flow management (ATFM). ATFM has the main objective to regulate the flow of aircraft as efficiently as possible to avoid congestion of certain control sectors. It is implemented for airspace where traffic demand at times exceeds the defined air traffic control capacity (ICAO Doc 4444 2016). The steps and means used are directed towards resolving the imbalance between supply and demand and towards providing adequate capacity at the required place and time. Supply and demand can be managed by imposing various restrictions on certain traffic flows (e.g., re-routing certain traffic flows, scheduling or rescheduling flights as appropriate, etc.). Also, supply can be increased by proper sector management (e.g., increasing the number of controllers working at the same time). Hazardous or restrictive meteorological conditions have a direct and immediate (negative) impact on ATFM by reducing airspace and aerodrome capacity. AFTM measures can be seen as strategic and pre-tactical, as they do not affect the current situation, but rather the near future.

Airspace management (ASM). ASM has the objective to manage airspace as efficiently as possible to satisfy its many civil and military users. This service is based on seamless and collaborative management of airspace configurations and concerns both the way airspace is allocated to its various users (using routes, zones, flight levels, etc.) and how it is structured to provide air traffic services. Moreover, ASM prevents mutual interference from all users of the airspace, facilitates air defence identification, and accommodates the flow of all air traffic safely.

4.2 Air traffic control jurisdiction

The territories over which certain air traffic control service provider has jurisdiction encompass portions of the airspace and aerodromes where air traffic services will be provided. This may include the complete airspace of a state or even a region. In certain cases, by mutual agreement, one state may delegate to another state the responsibility for establishing and providing air traffic services in the airspace over its territories. Moreover, in parts of the airspace over the oceans or in the airspace of undetermined sovereignty, air traffic services are generally provided by one or more states that accepted this responsibility based on regional air navigation agreements (ICAO Annex 11 2018). For example, Canada or the US control the eastern half of the Pacific in the northern hemisphere, while Russia, Japan, and China control the western side.

Once the ATM is established, the state concerned will decide whom to designate the authority for providing such services. The authority responsible for providing the services may be the state itself or a suitable agency (ICAO Annex 11 2018).

All commercial flights intending to operate in whole or in part within the designated airspace of a state are required to be approved by the state's ATS provider. Once the flight is approved, in most cases, it is controlled in whole or in part by that ATS provider (Babić and Netjasov 2016).

4.3 Air traffic control system capacity

The capacity of the air traffic control system is the ability to provide air navigation services to a certain volume of air traffic, taking into account the targeted level of safety and without imposing significant operational, economic, or environmental penalties under normal circumstances (Eurocontrol 2003). Among many capacity definitions, there are two that are most used:

- Sector capacity—the maximum number of flight entries in an hour that can safely be assigned to sector controllers;
- Declared sector capacity—the number of flight entries per hour, assessed by air navigation service provider and declared by the Central Flow Management Unit (CFMU),[1] so that the sector can be protected from overload.

Generally, the ATC capacity is expressed through the sustainable hourly traffic flow, which can be further converted into daily, monthly or annual values.

The capacity of an ATS system determines the available airspace capacity (complete airspace and by sectors) and the capacity of the divisions of the ATC (complete and by divisions). The capacity of an ATS system depends on many factors, including the ATS route structure, the navigation accuracy of the aircraft using the airspace, weather-related factors, and controller workload. The most common factors that are taken into account for assessing the ATC capacity values are (ICAO Doc 4444 2016):

(a) the level and type of ATS provided;

(b) the structural complexity of the control area, the control sector, or the aerodrome concerned;

(c) controller workload, including control and coordination tasks to be performed;

[1] CFMU within Eurocontrol plays a critical role in planning for and coordinating the response to contingency when the loss of service in one member state can have knock-on effects across the air traffic management (ATM) network.

(d) the types of communications, navigation, and surveillance systems in use, their degree of technical reliability and availability, as well as the availability of backup systems and/or procedures;
(e) availability of ATC systems providing controller support and alert functions and
(f) any other factor or element deemed relevant to controller workload.

The main task of ATC is to provide sufficient capacity to accept both normal and peak traffic levels. However, there are defined measures that can be implemented to increase/decrease the available capacity, while keeping the targeted safety levels. The reasons why the capacity should be changed can be numerous. In the cases of disturbances, such as any particular event that negatively impacts the declared capacity of an airspace or aerodrome, the capacity of the airspace or aerodrome concerned shall be reduced accordingly, for the required period (ICAO Doc 4444 2016). Moreover, in the areas where traffic demand varies significantly on a daily or periodic basis, the ATS system has defined facilities and procedures that are implemented accordingly to vary the number of operational sectors or working positions to meet the demand. This is particularly important in cases when the demand exceeds the available ATC capacity, and it is mandatory to ensure that safety is not jeopardized.

Flexible Use of Airspace (FUA). The increasing density of air traffic in the last decade has required new standards and procedures to efficiently accommodate the increasing air traffic, and, thus, the availability of airspace became even more important. According to ATM data, there are on average 26,000 flights daily in the European routes (FABEC 2014). Forecasts indicate that the volume of air transport will double by 2030, (Eurocontrol 2018), which makes traffic demands significantly larger than the capacity of airspace. Instead of being a purely civil or military property, airspace is now designated as a continuum, which is allocated flexibly and always temporarily, to civil and/or military users according to their requirements.

Flexible use of airspace (FUA) is an airspace management concept based on the principle that airspace should not be designated as purely civil or military, but rather as a continuum in which all user requirements are accommodated to the greatest possible extent to increase airspace capacity and to improve the efficiency and flexibility of aircraft operations (ICAO Doc 4444 2016). FUA means the availability of airspace for military purposes, which is temporarily segregated for reasons of flight safety, and flexible ATS routes for general air traffic. These routes supplement the existing ATS route network (they can be planned and used under specified conditions) and lead through temporary segregated airspace

(TSA). Of course, any necessary airspace segregation is temporary, based on real-time usage within a specific period.

This FUA Concept was introduced in March 1996 after its development by civil and military representatives of the European Civil Aviation Conference (ECAC) States together with representatives of aircraft operators. It facilitates direct civil and military coordination to avoid airspace segregation as much as feasible. Moreover, FUA enables continuous information flows between the three ASM levels and optimizes ATM network performances of both civil and military aviation. The latter is a real challenge due to the different needs of the users concerned, flight efficiency for civil aviation, and mission effectiveness for military aviation (ICAO Cir330 AN/189 2011).

The FUA concept has been developed at the three Levels of Airspace Management that correspond to Civil/Military coordination tasks. Each Airspace Management Level has an impact on the others:

- Level 1—strategic—definition of the national airspace policy and establishment of pre-determined airspace structures;
- Level 2—pre-tactical—day-to-day allocation of an airspace according to user requirements;
- Level 3—tactical—real-time use of airspace allowing safe Operational Air Traffic and General Air Traffic (OAT and GAT) operations.

The implementation of the FUA concept has already benefited both civil and military aviation concerning increased efficiency (reduction in distance, time, and fuel), the establishment of an enhanced ATS route network (increase in ATC capacity and reduction in delays to general air traffic), enhanced real-time civil/military coordination, as well as a reduction in airspace segregation needs.

4.4 Expression of vertical position of aircraft

The vertical position of the aircraft is the height at which the aircraft is at a particular moment of the flight. Depending on the plane that is taken as the basis for determining the vertical position of the aircraft, the height is defined in one of the following ways, Figure 4.2 (ICAO 8168 2018):

- *Elevation (E)*—The vertical distance of a level, a point, or an object considered as a point on the ground, measured from a mean sea level (MSL). The pressure set on the subscale of the altimeter so that the instrument indicates its height above MSL is known as **QNH pressure**. The altimeter will read runway elevation when the aircraft is on the runway.

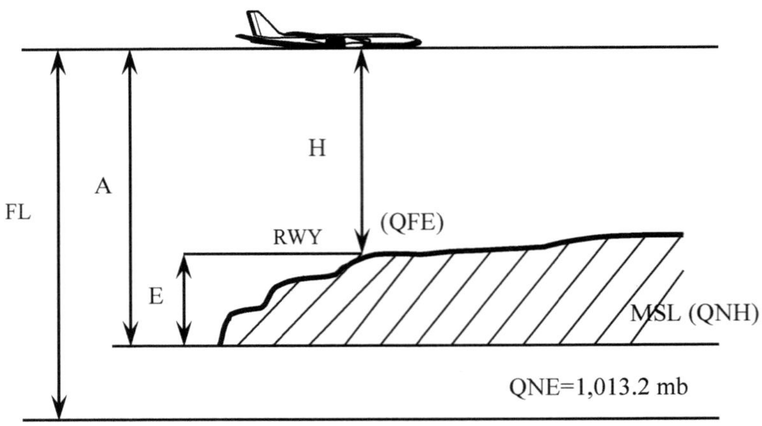

Figure 4.2. Vertical position of aircraft.
Source: Babić and Netjasov 2016

- *Height (H)*—The vertical distance of a level, a point, or an object considered as a point in the air, measured from a specified datum (e.g., runway). The pressure set on the subscale of the altimeter so that the instrument indicates its height above the reference elevation being used is known as **QFE pressure**. QFE is the air pressure at the current ground level. It is used to cause the altimeter to register height above the ground.
- *Altitude (A)*—The vertical distance of an object in the air measured from MSL. It is determined by using **QNH pressure.**
- *Flight Level (FL)*—A surface of constant atmospheric pressure which is related to a specific pressure datum, 1,013.2 hectopascals (hPa) at MSL at standard air pressure. The pressure set on the subscale of the altimeter so that the instrument indicates its height above the reference plane being used is known as **QNE pressure**. Flight levels are separated from each other at discrete pressure intervals of 18 hPa (this corresponds to a change in height of 500 ft)[2] expressed in hundreds of feet. Flight levels are generally expressed as FLxxx, where the xxx is a two- or three-digit number indicating the pressure altitude in units of 100 ft. For example, FL310 corresponds to an altitude of 31,000 ft above MSL when the pressure at sea level is 1,013.2 mb.

An altimeter is an instrument used to measure the altitude of an object above a fixed level. It measures ambient air pressure, which decreases with increasing altitude (approximately 100 hPa at 800 m, i.e., 1 mmHg

[2] The primary unit of measurement of altitude and elevation or height in aviation is the foot [ft], (1 ft = 0.31 m).

of column at 1,000 ft above sea level). The altimeter then calculates and displays the corresponding altitude. It is essentially a calibrated barometer that displays an altitude—a barometric altimeter.

If the barometric altimeter is calibrated to the standard atmosphere and set to:

- QNH pressure, it displays the altitude/elevation;
- QFE pressure, it displays the height relative to the reference plane on earth;
- QNE pressure (1,013.25 hPa) displays the flight level.

Due to natural local variations in atmospheric air pressure and variations over time (daily and seasonal), the vertical position of an aircraft shown by a barometric altimeter will also differ depending on the location and time where it is measured, as well as about which plane it is determined. Therefore, a surface of constant atmospheric pressure was introduced (1,013.25 hPa).

If altimeters on different aircraft are not calibrated consistently, then it could happen that these two aircraft could be flying at the same altitude even though their altimeters show that they are at different altitudes. This issue is solved by using flight levels, i.e., defining altitudes based on standard air pressure at sea level. Thus, all aircraft operating on flight levels must be calibrated to this setting (QNE pressure) regardless of the actual sea level pressure. To display true altitude above sea level, the pilot would need to calibrate the altimeter according to the local air pressure at sea level (QNH pressure).

The general procedure for flights in the vicinity of aerodromes and within terminal control areas is that the vertical position of aircraft is expressed in terms of altitudes at or below the transition altitude, while the vertical position of aircraft is expressed in terms of flight levels at or above the transition level.

The transition altitude (TA) is the altitude at or below which the vertical position of an aircraft is expressed in terms of altitude, which is determined from an altimeter set to QNH pressure (ICAO 8168 2018). When operating at or below the TA, aircraft altimeters are usually set to show the altitude above sea level. Above the TA, the aircraft altimeter pressure setting is normally adjusted to the standard pressure setting of 1,013.25 hPa and the aircraft altitude will be expressed as the flight level. TA is the predetermined altitude for each airport by the state in which the aerodrome is located, due to the local variations in atmospheric air pressure. Where two or more closely spaced aerodromes are located so that coordinated procedures are required, a common transition altitude is established (ICAO 8168 2018).

a)

b)

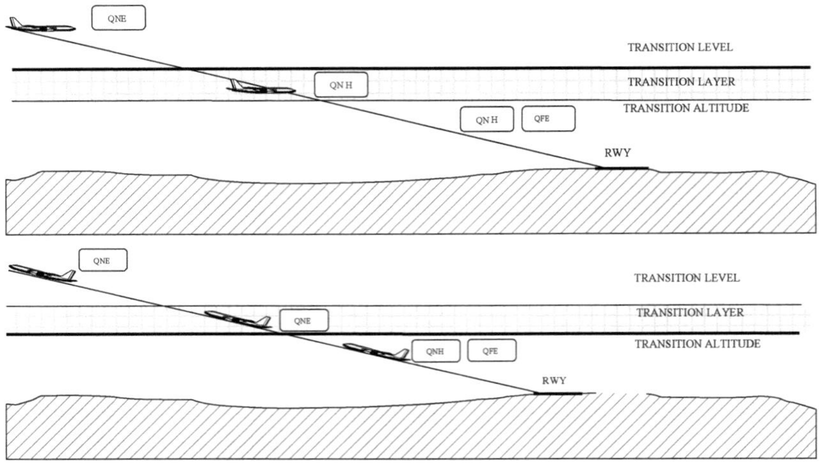

Figure 4.3. Altimeter adjustment procedure during (a) descending and landing (b) take-off and climbing.
Source: Babić and Netjasov 2016

The transition level (TL) is the lowest flight level available for use above the transition altitude (ICAO 8168 2018). At TL and above, it is flown with an altimeter set to QNE pressure. TL is usually passed to pilots together with approach and landing clearances.

The transition layer (TLY) is the airspace between the transition altitude and the transition level. During landing, when an aircraft is approaching the transition level, the pilot sets the altimeter to QNH pressure. With the altimeter that now shows altitude, the aircraft descends through a transition layer and a transition altitude. The altimeter remains set to QNH pressure until the landing is completed (Figure 4.3a). During take-off, the opposite procedure is applied, when the aircraft reaches the transition altitude, the pilot adjusts the altimeter to the QNE pressure and thus flies through the transition layer (Figure 4.3b).

4.5 Airspace division

The airspace of a state or region includes the space above a particular national territory (land and coastal sea), from the surface and is usually unlimited in height, and coincides with the borders of the state or the borders of the observed region on land and sea. It can be divided horizontally and vertically, and further into smaller units of space (Babić and Netjasov 2016). The categories and types of airspace are dictated by the complexity or density of aircraft movements, nature of the operations conducted within the airspace, the level of safety required, and national and public interest.

4.5.1 Horizontal division of airspace

Horizontally, airspace is divided into the following sections:

1. Controlled airspace;
2. Uncontrolled airspace;
3. Special Use Airspace.

Controlled airspace. Controlled airspace is the particular portion of the airspace or at particular aerodromes within which air traffic control services will be provided. It is actively monitored and managed by air traffic controllers. To enter controlled airspace, an aircraft must first gain clearance from the air traffic controller. It consists of the following areas:

(a) *Flight Information Region (FIR)*—Those portions of the airspace where it is determined that flight information service and alerting service will be provided to each aircraft flying within it (ICAO Annex 11 2018). FIR airspace boundaries usually follow state borders, Figure 4.4. All airspace around the world is divided into Flight Information Regions (FIRs). Airspace over the ocean is typically divided into two or more FIRs and delegated to controlling authorities within countries that border it. In some cases, FIRs are split vertically into lower and upper sections. The lower section remains referred to as an FIR, but the upper portion is referred to as an Upper Information Region (UIR). FIRs vary in size. Smaller countries may have one FIR/UIR in the airspace above them and larger countries may have several. Airspace within an FIR (and UIR) is usually divided into pieces that vary in function, size, and classification. Classifications determine the rules for flying within a piece of airspace and whether it is controlled or uncontrolled.

(b) *Control area (CTA)*—The portion of the airspace where it is determined that air traffic control service is provided to each Instrument Flight Rules (IFR) flight within it. It exists in the vicinity of the airport and extends upwards from a specified limit above the earth Figure 4.5. A lower limit of CTA should be established at a height above the ground or water of not less than 200 m (700 ft). It has a specified lower level and a specified upper level (Upper Control area—UTA). This area can be divided into more sectors (ICAO Annex 11 2018).

(c) *Terminal Control Area (TMA)*—It is also known as a terminal manoeuvring area and it represents controlled airspace surrounding a major airport with a high volume of traffic. TMA is usually established at the confluence of ATS routes in the vicinity of one or more major aerodromes, Figure 4.6 (ICAO Annex 11 2018). It extends from the surface or the upper limit of the controlled zone to the lower limit of CTA/UTA.

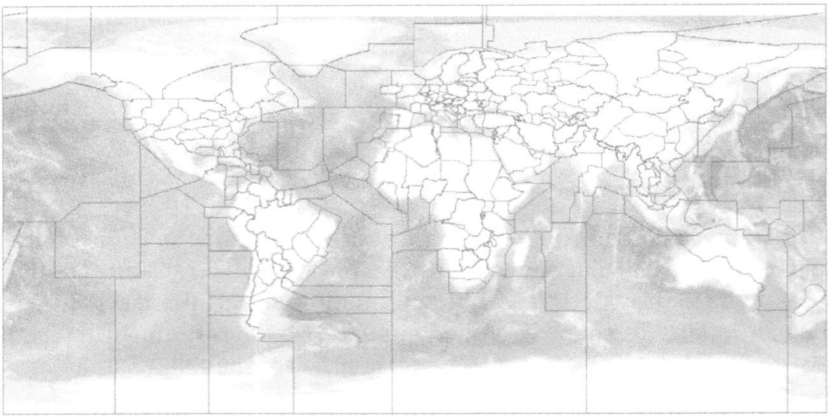

Sources: Esri, USGS, NOAA, Esri, HERE, DeLorme, MapmyIndia, © OpenStreetMap contributors, and the GIS user community

Figure 4.4. Flight Information Regions (FIRs) around the world.

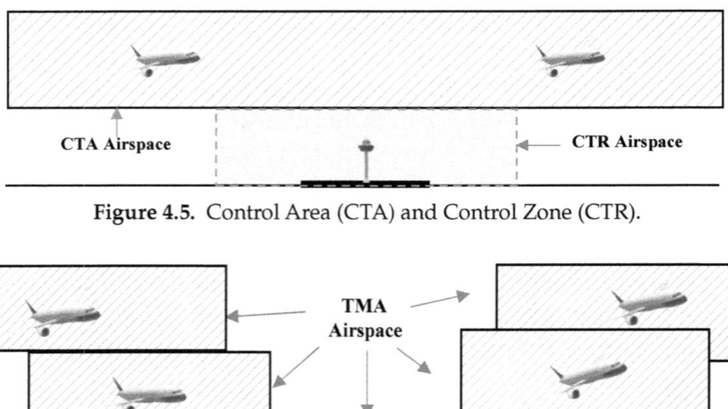

Figure 4.5. Control Area (CTA) and Control Zone (CTR).

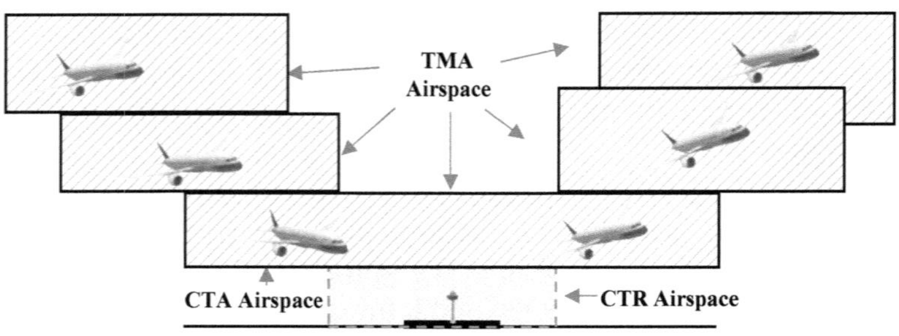

Figure 4.6. Terminal Control Area (TMA) and Control Zone (CTR).

(d) *Control Zone (CTR)*—is controlled airspace extending upwards from the surface of the earth to a specified upper limit (to the CTA or TMA lower limit), established to protect air traffic operating to and from that airport. The lateral limits of CTR encompass those portions of the airspace which are not within control areas containing the paths of IFR flights arriving at and departing from aerodromes, Figures 4.5 and 4.6. This airspace is usually under the control of a tower controller (ICAO Annex 11 2018).

(e) *Airways (AWY)*—is a control area or portion thereof established in the form of a corridor (ICAO Annex 11 2018). Airways are defined segments within a specific altitude block, corridor width of 10 nautical miles (NM), and between points in airspace based on (a) geographic coordinates named fix(es) for a satellites navigation system, or (b) radio navigational aids (navaids) (such as VORs or NDBs) or (c) the intersection of specific radials of two navaids. They link major airports giving protection to IFR flights during the climb and descent phases. Public transport can be performed exclusively within airways. Recently, a new concept has been implemented known as *free route airspace* where airspace users may plan flight routes freely between designated points (see 4.11).

Uncontrolled airspace. Uncontrolled airspace is not supervised by an air traffic control provider; thus no clearance is required to operate within this airspace. It includes all the airspace of a country or region that is outside and below the lower limits of controlled airspace. Uncontrolled airspace is also called the free flight layer and extends from the surface up to 900 m in absolute height or 450 m above the ground, whichever is higher (Babić and Netjasov 2016).

Uncontrolled airspace is mostly used by light aircraft and helicopters that operate outside or underneath the controlled airspace. In uncontrolled airspace, pilots are often not visible to air traffic controllers but must still follow visual flight rules (VFR) or instrument flight rules (IFR). In uncontrolled airspace controllers do not provide separation but provide a flight information service and traffic information service to aircraft flying on instrument flight rules and upon request to aircraft flying on visual flight rules.

Classification of airspaces. The airspace is further divided into different classes, where internationally agreed rules for visual flight and instrument flight apply. Depending on how far and how high an aircraft wants to fly, it will pass through different classes of airspace, in which different rules apply. Controlled airspace consists of (ICAO Annex 11 2018):

a. **Class A** airspace is the airspace where all operations are conducted under instrument flight rules and all flights are provided with air traffic control service, Figure 4.7. Aircraft are separated from all other traffic and the users of this airspace are mainly major airlines and business jets.

b. **Class B** airspace is the airspace surrounding an airport area in terms of airport operations or passenger enplanements. The configuration of each Class B airspace area is individually tailored, consists of a surface area and two or more layers, and is designed to contain all published instrument procedures once an aircraft enters the airspace. IFR and

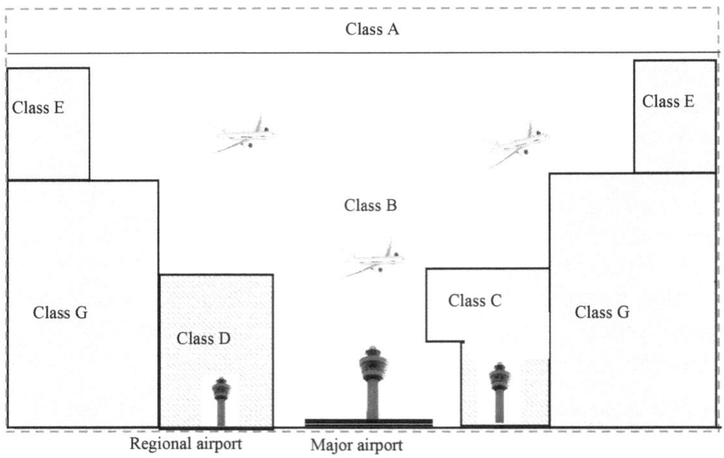

Figure 4.7. Airspace classification.

VFR flights are permitted. All aircraft entering Class B airspace must obtain ATC clearance before entry and must be prepared for denial of clearance.

c. **Class C** airspace is the airspace surrounding those airports that have an operational control tower, are serviced by a radar approach control and have a certain number of IFR operations or passenger enplanements (a moderate traffic level) Figure 4.7. IFR and VFR flights are permitted, all flights are provided with air traffic control service and IFR flights are separated from other IFR flights and VFR flights. The configuration of each Class C area is individually tailored. Each aircraft must establish two-way radio communications with the ATC facility providing air traffic services before entering the airspace and thereafter must maintain those communications while within the airspace.

d. **Class D** airspace is the airspace surrounding those smaller airports that have an operational control tower. The configuration of each Class D airspace area is individually tailored and, when instrument procedures are published, the airspace is normally designed to contain the procedures. IFR and VFR flights are permitted and all flights are provided with air traffic control service Figure 4.7. Each aircraft must establish two-way radio communications with the ATC facility providing air traffic services before entering the airspace and must thereafter maintain those communications while in the airspace.

e. **Class E** airspace is the controlled airspace not classified as Class A, B, C, or D airspace, Figure 4.7. This portion of airspace is used for safe control and separation of aircraft during IFR operations. IFR and VFR flights are permitted, but only IFR flights are provided with air traffic control service and are separated from other IFR flights.

f. **Class G** airspace is the designated uncontrolled airspace. Class G airspace extends from the surface to the base of the overlying Class E airspace. IFR and VFR flights are permitted and receive flight information service if requested Figure 4.7. ATC has no authority or responsibility to control air traffic herein.

Special Use Airspace. Airspace may be further subdivided into a variety of areas and zones, including those where there are either restriction on flying activities or complete prohibition of flying activities. Special use airspace is the designation for airspace in which certain activities must be limited, or where limitations may be imposed on aircraft operations that are not part of these activities. Certain special use airspace areas can create limitations on the mixed use of airspace. The special use airspace depicted on instrument charts includes the area name or number, effective altitude, time and weather conditions of operation, the controlling agency, and the chart panel location. Special use airspace usually consists of the following (ICAO Annex 2 2005):

- *Prohibited area* contains airspace of defined dimensions within which the flight of aircraft is prohibited. Such area is established for security or other reasons associated with the national welfare (e.g., airspace around nuclear power plants, petrochemical plants, etc.). The prohibited area is charted as a "P" followed by a number (e.g., P-40).
- *Restricted area* is an area of defined dimensions, above the land areas or territorial waters of a state, within which the flight of aircraft is restricted with certain specified conditions. In restricted airspace, aircraft movements are reduced to those with certain specified permissions. Examples include airspace around military installations, high-density flying operations at an air show, or volcanic activity. The restricted area is charted with an "R" followed by a number (e.g., R-4401) and is depicted on the en-route chart appropriate for use at the altitude or FL being flown.
- *Danger area* is an airspace of defined dimensions within which activities dangerous to the flight of aircraft may exist at specified times ICAO Annex 11 2018. Generally, dangerous areas are operated by military authorities (e.g., military exercises involving live firing, parachute dropping, violent and unpredictable aircraft manoeuvres, etc.). Vertical and lateral dimensions of danger areas are publicised in national Aeronautical Information Publications (AIPs) together with the hours of operation where applicable. The danger area is charted with a "D" followed by a number (e.g., D-121).

- *Temporary segregated airspace (TSA)* is an area that is temporarily reserved and allocated for the exclusive use of a specific user during a determined period. TSA is needed for specific military training operations.

4.5.2 Vertical division of airspace

Airspace is split vertically into several sections between the surface or sea level and a certain flight level (Babić and Netjasov 2016):

1. *Free flight layer*—uncontrolled airspace used by VFR flights that are not controlled by the air traffic control provider. It extends from the surface up to 900 m in absolute height or 450 m above the ground, whichever is greater.
2. *Lower airspace*—extends from the upper limit of the terminal control area to the FL285. The highest usable flight level is FL280.
3. *Upper airspace*—extends from the upper limit of the lower airspace (FL285) to the FL450 that represents the highest usable flight level. The lowest usable flight level is FL290.
4. The airspace where traffic demand is high can be divided into additional layers: *Middle, Top Upper,* and *Super Top Upper* airspace.

4.6 Meteorological conditions and rules of the air

Flight conditions refer to the state of the atmosphere expressed in terms of visibility, distance from cloud, and ceiling. There are two types of meteorological conditions in aviation (ICAO Annex 2 2005):

- *Visual Meteorological Conditions (VMC)*—Meteorological conditions expressed in terms of visibility, distance from cloud, and ceiling, equal to or better than the specified minima;
- *Instrumental Meteorological Conditions (IMC)*—Meteorological conditions expressed in terms of visibility, distance from cloud, and ceiling, less than the minima specified for visual meteorological conditions.

Depending on the flight conditions, a flight must be conducted in accordance with the following rules of the air (ICAO Annex 2 2005):

- *Visual Flight Rules (VFR)*—refers to a set of rules created for flight in VMC or visual meteorological conditions;
- *Instrumental Flight Rules (IFR)*—a set of rules that govern aircraft that fly in IMC, or instrument meteorological conditions.

During VFR operations, air traffic control is not responsible for keeping VFR aircraft separated. Thus, the responsibility for traffic separation belongs solely to the pilot, i.e., the pilots need to be able to see in front of and around their aircraft while in the air. This is why VFR rules also cover visibility requirements and cloud clearance criteria required to fly with visual reference to the ground and/or horizon. These visibility and cloud clearance requirements vary depending on the type of airspace the pilot is flying in. However, they exist to ensure that pilots flying VFR do not get spatially disoriented, which may lead to both losses of control and controlled flight into terrain. Under VFR rules, the pilot is also responsible for being able to see other aircraft, avoiding collisions, and maintaining minimum horizontal separation of 3 NM to 5 NM (ICAO Doc 4444 2016).

IFR flying means flying in the clouds. More specifically, IFR flying means navigating a plane through bad weather conditions, heavy clouds, or under the darkness of night, leveraging cockpit instruments as altimeters, GPS systems, and vertical speed indicators. It is called instrument flight because the pilot navigates only by reference to the instruments in the aircraft cockpit. A pilot may select to fly with instrument flight rules in visual meteorological conditions or may be required to do so by the appropriate ATS authority. The pilots of IFR flights are obligated to maintain continuous voice communication and establish two-way communication with the corresponding air traffic control unit. Within the controlled airspace, all commercial aircraft are obligated to fly under IFR, regardless of whether they operate during the day or night and in any visibility conditions.

4.6.1 Separation standards

Separation standards refer to the minimum distance between the aircraft operating in controlled airspace and at airports with an operational control tower, that must be applied (ICAO Doc 4444 2016). These separation standards are important to facilitate the safe navigation of aircraft in controlled airspace. Depending on whether a flight is IFR or VFR, different separation standards are applied. Air traffic controllers are responsible for keeping the aircraft separated vertically or horizontally. Observance of these standards ensures safe separation from the ground, from other aircraft, and protected airspace.

In certain situations, separation standards serve to reduce exposure to wake vortex turbulence. Wake vortex turbulence will be explained later in detail in a subsequent section.

National separation standards are based on the provisions of ICAO Doc 4444 (Procedures for Air Traffic Management 2016). If there are any differences from these standards, they are published in national AIPs.

Aircraft vertical separation. Vertical separation is obtained by requiring aircraft to use prescribed altimeter setting procedures within designated airspace and to operate at different levels expressed in terms of flight levels or altitudes. ICAO specifies the minimum vertical separation for IFR flight as 1,000 ft (300 m) below FL290 and 2,000 ft (600 m) above FL290 (ICAO Doc 4444 2016). In the portions of airspace where Reduced Vertical Separation Minima (RVSM) are applied, the vertical separation minimum differs. Most national authorities in the world follow this rule, but they may specify a different level at which the rule changes (a value of less than 600 m (2,000 ft) but not less than 300 m (1,000 ft) to be used by aircraft operating above FL290 within designated portions of the airspace).

The difference in the prescribed vertical separation minimum for a certain flight level was necessary to introduce due to the reduction in pressure-sensing accuracy of barometric altimeters with increasing altitude. In 1966, an increased vertical separation minimum of 600 m (2,000 ft) was established above FL290 on a global basis. However, in the 1980s, ICAO and worldwide navigation offices started a study about the feasibility of reducing the vertical separation minima above FL290 from 2,000 ft to 1,000 ft. In the period between 1997 and 2005, RVSM was implemented in all of Europe, North America, South America, North Africa, Southeast Asia, and over the North Atlantic, the South Atlantic, and the Pacific Oceans. Soon after 2011, nearly all worldwide authorities have chosen RVSM, including China, Russia, and Africa.

RVSM is the reduction of the vertical separation minimum to 300 m or 1,000 ft between FL290 and FL410:

- 300 m or 1,000 ft below FL410;
- 600 m or 2,000 ft above FL410.

The principal benefits of the reduced vertical separation minimum were that allowing aircraft to operate closer to the optimum flight level (fuel economy) and increasing airspace capacity. For an airline operating in RVSM airspace, its aircraft must be equipped with (ICAO Doc 9574 2012):

- Two independent altitude measurement systems (altimeters) using independent pressure sensors;
- An altitude-deviation alerting system;
- An autopilot fitted with an altitude hold function;
- A secondary surveillance radar (SSR) transponder with altitude reporting synchronized with the altimeter being the source for the autopilot.

For aircraft to be safely separated during the flight, a semi-circular flight rule system is applied for vertical separation (Babić and Netjasov 2016). According to this system, all IFR flights flying the magnetic track between 000°–179° (eastbound) are obligated to use odd flight levels (e.g., FL250, FL270, etc.), while all IFR flights flying the magnetic track between 180°–359° (westbound) are obligated to use even flight levels (e.g., FL260, FL280, etc.) (Figure 4.8) (Babić and Netjasov 2016).

Aircraft horizontal separation. The horizontal separation between any two aircraft must be applied when they are separated by less than the vertical separation minimum. Lateral separation minima are usually based upon the position of an aircraft as determined by visual observation, by use of navigation aids, or by use of area navigation equipment.

The longitudinal separation must be applied in the case if two aircraft are not laterally separated and following the same or diverging tracks within 45 degrees of each other. This separation may be achieved by requiring aircraft to make position reports and comparing the time of their reports and by application of speed control.

When surveillance systems are used the minimum separation prescribed by ICAO Doc 4444 is 5 nm (unless otherwise stated by the appropriate ATS authority).

Wake Vortex Turbulence. Wake vortex, also known as wake turbulence, is defined as turbulence that is generated by the passage of an aircraft in flight. It can arise in several ways, e.g., it will be generated when the nose landing gear of an aircraft lifts off the runway on take-off, as in the case when the nose landing gear touches the ground during landing. However, potentially hazardous wake turbulence of an aircraft in flight is caused by wing tip vortices. Wake vortices are present behind every aircraft, they decompose quite slowly and are most dangerous to the following aircraft for several minutes after they have been generated (ICAO Doc 9426 1984). They are particularly severe when generated by large and wide-bodied jet aircraft.

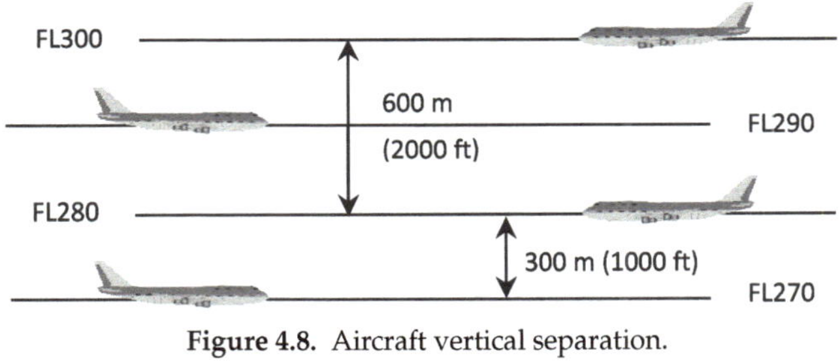

Figure 4.8. Aircraft vertical separation.

The aircraft wake turbulence is the by-product of lift. Lift occurs as a consequence of a pressure differential over the wing surface. The lowest pressure occurs over the upper wing surface and the highest pressure under the wing. This pressure differential results in the roll up of the airflow aft of the wing causing the creation of swirling air masses trailing downstream of the wing tips. The generated wake consists of two counter-rotating cylindrical vortices trailing aft from the aircraft (Figure 4.9).

Wake vortex is generated by an aircraft in flight and its strength is governed by the weight, speed, and shape of the wing of the generating aircraft. Its strength increases proportionally to weight and reaches maximum strength when the generating aircraft is large and wide-bodied, in a clean configuration, and is flying slowly. Vortices generally persevere for between one and three minutes. They are most persistent in stable air conditions with low wind speeds.

The potential for hazardous wake vortex turbulence is greatest where aircraft follow the same tracks. This situation is mostly encountered close to the ground in the vicinity of airports during the take-off, initial climb, final approach, and landing phases of flight. Wake vortex may cause sudden uncommanded roll moments to smaller aircraft by the effect of larger ones, which, in extreme cases, can be beyond the power of the flying controls or the prevailing response of the pilot to counteract. The three basic effects of wake turbulence on the following aircraft are induced roll, loss of height or rate of climb, and possible structural stress. If the generating aircraft is climbing or descending rapidly (greater than 1,000 feet per minute (fpm)), then a significant wake vortex may persist across several flight levels.

If the wake vortex has greater longevity at higher cruise altitudes this can lead to encounters at much greater in-track separation than

Figure 4.9. Wake vortex turbulence.

the normally applied ATC separation minima to IFR flights. Thus, the prescribed separation for aircraft receiving an ATC control service does not necessarily provide sufficient separation to remove the risk of loss of control.

Wake turbulence is invisible, thus, its presence and location cannot be determined with precision. However, when an en-route air traffic controllers identify a traffic proximity situation with risk of a potentially hazardous wake encounter, they may provide traffic information to the trailing aircraft, including a caution for potential wake turbulence and when possible, may propose a change of lateral or vertical flight path, as appropriate.

In addition to traffic separation standards applied to aircraft on approach to or taking off from the same or closely spaced runways, more restrictive separation minima are defined (see ICAO Doc 9426) to ensure that the following aircraft, especially smaller ones, are not hazarded by the effects of wake turbulence generated by the preceding aircraft. Aircraft are categorised according to their certificated Maximum Take-Off Mass (MT/OM) and minimum separation times or distances are applied to aircraft in sequence so that aircraft following a higher MT/OM aircraft are given greater spacing than those following the same or lower MT/OM category.

4.6.2 Right-of-way

An aircraft that has the right-of-way has an advantage in flight over other aircraft, due to its characteristics and performance. In that case, the aircraft that has the right-of-way maintains its heading and speed in respect to other aircraft. In aviation, the right-of-way is as follows (ICAO Annex 2 2005):

- power-driven heavier-than-air aircraft must give way to airships, gliders, balloons, and to aircraft which are seen to be towing other aircraft or objects;
- airships shall give way to gliders and balloons;
- aircraft with piston propulsion groups are removed and priority is given to aircraft with jet propulsion groups;
- gliders give way to balloons and aircraft towing other aircraft or objects, etc.

There are additional right-of-way rules applied when two aircraft meet in flight. One of them refers to the case when two aircraft with approximately the same landing time are approaching an aerodrome for landing. The right of priority is given to the aircraft that has a jet propulsion group in relation to the aircraft with a piston propulsion group. If both aircraft have the same propulsion group, the aircraft with a larger number of engines has the advantage. Also, when two aircraft approach the airport at the same time, aircraft at the higher level must give way to aircraft at the lower level.

Moreover, the right-of-way rules applied at civil airports, in respect to landing and take-off, which are defined due to the type of the aircraft and purpose of flight, are in the following order (ICAO Annex 2 2005):

- emergency aircraft (fuel shortage, failure of an on-board system, etc.) has the right-of-way in respect to all other aircraft;
- aircraft with sick or wounded passengers;
- search and rescue aircraft;
- aircraft carrying senior state officials with VIP status;
- aircraft performing scheduled and non-scheduled traffic;
- other aircraft, etc.

Also, there are rules related to giving way to vehicles moving on the manoeuvring surfaces of the airport. Based on these rules, a vehicle that pulls an aircraft on the manoeuvring surfaces of an airport has an advantage over aircraft that taxis and other vehicles. Also, a taxiing aircraft has the right-of-way over vehicles that do not tow other aircraft and other vehicles. An aircraft that is taking off or landing has the right-of-way over all other aircraft and vehicles moving on the manoeuvring area of an aerodrome. An aircraft that is taking off gives priority to a landing aircraft.

When two aircraft fly towards each other and are at the same altitude, both pilots are required to turn to the right in a timely manner (moment t_1), so that at the time of passing (moment t_2) the aircraft are separated by a minimum of 300 m (Figure 4.10a) (ICAO Annex 2 2005). Approaching head-on means a situation when the trajectories of two aircraft differ by β=180° ± 20° (Figure 4.10b).

In case of converging the paths of aircraft flying at the same altitude, the pilot who sees another aircraft from his/her right side is obliged to turn to his right side in time (moment t_1), so that at the moment of crossing (moment t_2) the aircraft are separated by at least 300 m (Figure 4.11a) (ICAO Annex 2 2005). A converging is a situation where the paths of two aircraft intersect at an angle of β = 70° up to β = 160° (Figure 4.11b).

The last case is when an aircraft overtakes another aircraft at the same altitude when climbing or sinking. An overtaking aircraft is an aircraft that approaches another from the rear on a line forming an angle of less than 70 degrees (β = 70°), Figure 4.12b, with the plane of symmetry of the latter (ICAO Annex 2 2005). An aircraft that is being overtaken has the right-of-way, and the overtaking aircraft, whether climbing, descending, or in horizontal flight, is obliged to turn

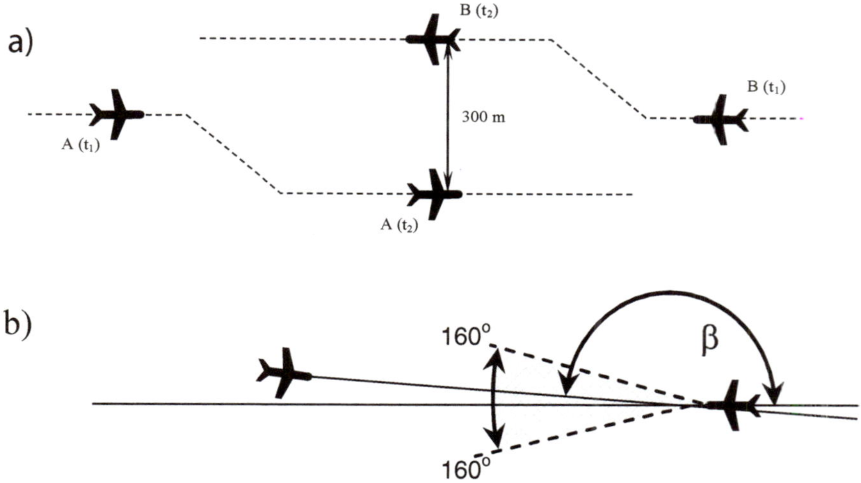

Figure 4.10. The right-of-way when approaching head-on.
Source: Babić and Netjasov 2016

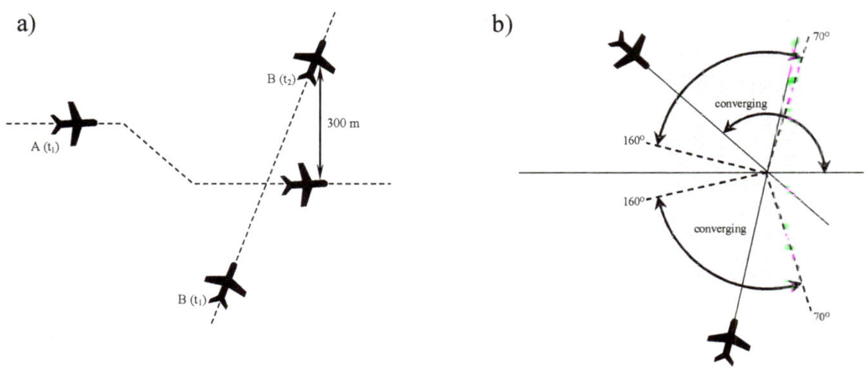

Figure 4.11. The right-of-way when converging.
Source: Babić and Netjasov 2016

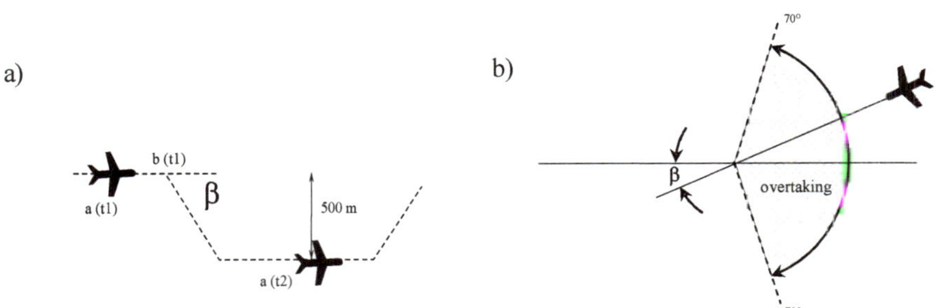

Figure 4.12. The right-of-way when overtaking.
Source: Babić and Netjasov 2016

to its right (moment t_1) so that at the moment of overtaking a minimum lateral distance of 500 m is achieved (moment t_2) (Figure 4.12a).

4.7 Flight plan

An ATC flight plan is a document that provides specified information to air traffic service units relative to an intended flight or portion of a flight of an aircraft (ICAO Annex 2 2015).

This plan is submitted before operating any flight or portion thereof to be provided with air traffic control service and it outlines the route the aircraft will take and alternative plans in the event of an emergency or a weather-related problem. An airline files a flight plan with the air traffic control so every controller who deals with the flight on its journey is aware of its details and route. A flight plan must be submitted at least sixty minutes before departure, or, if submitted during the flight, at a time which will ensure its receipt by the appropriate air traffic services unit at least ten minutes before the aircraft is estimated to reach the intended point of entry into a control area or advisory area or the point of crossing an airway or advisory route (ICAO Annex 2 2015). Once the flight plan is approved, the pilot is ready to contact the ground controller for taxi instructions before the departure.

A flight plan contains the following information relevant to the flight for the appropriate ATS authority (Figure 4.13):

1. Aircraft identification;
2. Flight rules and type of flight;
3. Number and type(s) of aircraft and wake turbulence category;
4. Equipment;
5. Departure aerodrome;
6. Estimated off-block time;
7. Cruising speed(s);
8. Cruising level(s);
9. Route to be followed;
10. Destination aerodrome and the total estimated elapsed time;
11. Alternate aerodrome(s);
12. Fuel endurance;
13. Total number of persons on board;
14. Emergency and survival equipment;
15. Other information.

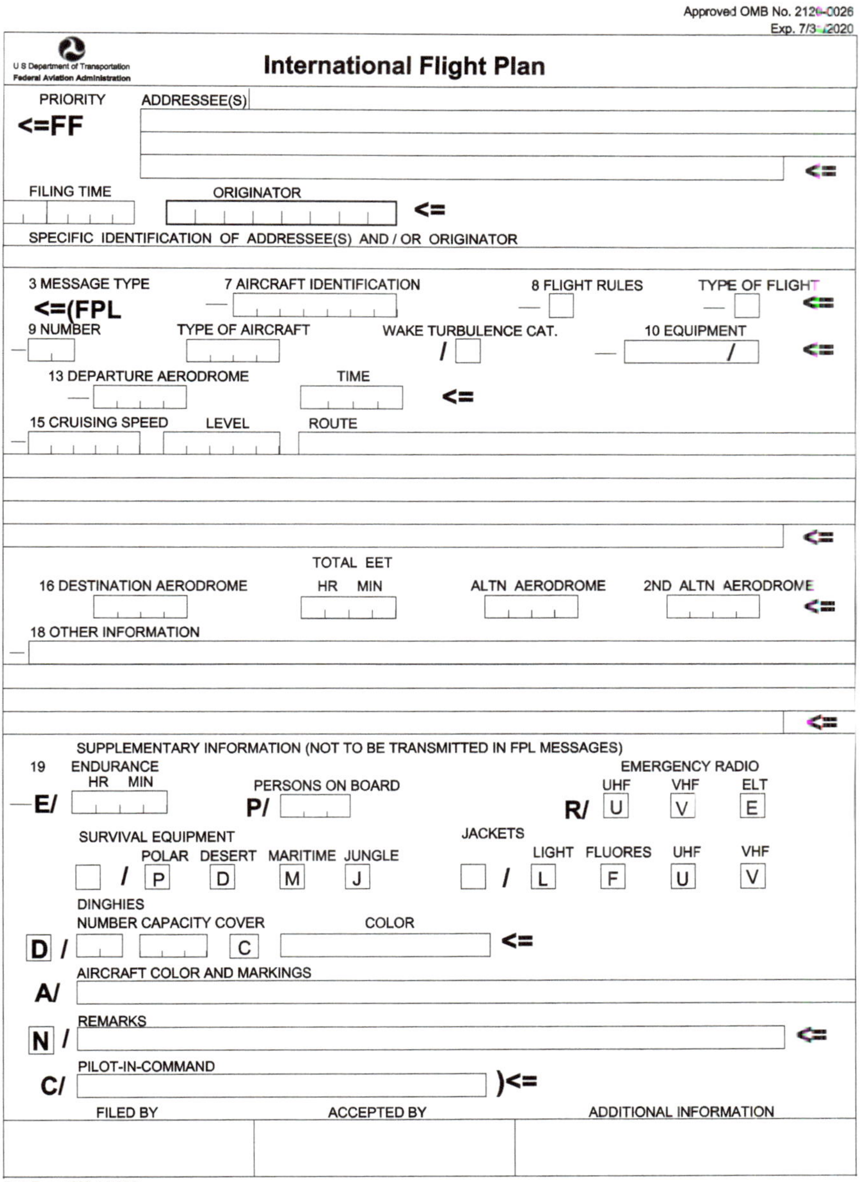
Approved OMB No. 2120-0026
Exp. 7/31/2020

U S Department of Transportation
Federal Aviation Administration

International Flight Plan

PRIORITY <=FF
ADDRESSEE(S) <=

FILING TIME
ORIGINATOR <=

SPECIFIC IDENTIFICATION OF ADDRESSEE(S) AND / OR ORIGINATOR

3 MESSAGE TYPE <=(FPL
7 AIRCRAFT IDENTIFICATION
8 FLIGHT RULES
TYPE OF FLIGHT <=

9 NUMBER
TYPE OF AIRCRAFT
WAKE TURBULENCE CAT. /
10 EQUIPMENT / <=

13 DEPARTURE AERODROME
TIME <=

15 CRUISING SPEED
LEVEL
ROUTE <=

16 DESTINATION AERODROME
TOTAL EET HR MIN
ALTN AERODROME
2ND ALTN AERODROME <=

18 OTHER INFORMATION <=

SUPPLEMENTARY INFORMATION (NOT TO BE TRANSMITTED IN FPL MESSAGES)

19 ENDURANCE HR MIN — E/
PERSONS ON BOARD P/
EMERGENCY RADIO UHF VHF ELT R/ U V E

SURVIVAL EQUIPMENT POLAR DESERT MARITIME JUNGLE / P D M J
JACKETS LIGHT FLUORES UHF VHF / L F U V

DINGHIES NUMBER CAPACITY COVER COLOR D / C <=

AIRCRAFT COLOR AND MARKINGS A/

REMARKS N / <=

PILOT-IN-COMMAND C/)<=

FILED BY | ACCEPTED BY | ADDITIONAL INFORMATION

FAA Form 7233-4 (7/15)

Figure 4.13. International Flight Plan Form.
* All clock times are expressed with 4 figures in UTC reference time.

Although flight plans are compulsory when flying in controlled airspace, there are a few exceptions when pilots may deviate from the approved flight plan. The pilot in command may deviate from the approved clearance if the amended clearance is obtained, an emergency exists, or the deviation is in response to a traffic alert and collision avoidance system resolution advisory. However, except in Class A airspace, a pilot may cancel an IFR flight plan if the operation is being conducted in VFR weather conditions. When a pilot is uncertain of an ATC clearance, that pilot must immediately request clarification from the ATC.

4.8 Surveillance systems in air traffic control

There are two systems applied for the provision of air traffic control service: the procedural control system and the radar control system. Regardless of whether procedural or radar control methods are used, air traffic controllers are trained to work in both systems. This means that controllers monitor traffic situations, detect and solve conflicts by providing separation, and ensure an orderly flow of air traffic.

Procedural control system is a method for the provision of air traffic control service without the use of the radar (ICAO Doc 4444 2016). This type of system is still in use in those areas that are sparsely populated, where radars are not available or impossible to install (in the airspace above the ocean), and in areas where installation of radars is unprofitable. Also, the procedural system is still applied at smaller airports (TMA and CTR), where traffic levels are such that they do not justify the installation and maintenance of a surveillance system. The procedural system is used as a backup solution in case of complete failure of all surveillance-based systems (ICAO Doc 4444 2016).

In the procedural system, standard vertical separation is 1,000 ft at or below FL280 (or at or below FL410 if RVSM is applied) and 2,000 ft above that level. For horizontal separation of two aircraft located at the same altitude, the separation minima are based on the distance between two aircraft and are expressed and applied in minutes or nautical miles. Different minima may be applied based on aircraft speed (the faster the former aircraft, the smaller the separation minimum) and navigation equipment (if aircraft position can be determined more accurately, smaller minima apply). For aircraft flying successively one after the other, a separation of at least 10 minutes shall be applied.

Data to be used by controllers in the procedural system are obtained from the following data sources:

- Data on aircraft intentions and other data are obtained from the flight plan;
- Data for coordination and transfer of air traffic control is received by telephone or intercom;

- Meteorological data in the form of reports are received from meteorological stations, forecasts are received from meteorological stations at the airport, and information and warnings are received from competent meteorological observation centres;
- Information on the position of the aircraft in the airspace is obtained from either pilot reports or visual observations, the latter being applicable in the CTR or on the manoeuvring area through tower and ground controllers.

From flight plan data (aircraft identification, altitude, speed, aircraft type, take-off, and landing aerodrome, route, etc.), air traffic controllers complete **a flight progress strip** for each aircraft that will pass through their controlled airspace (ICAO Doc 4444 2016). Controllers use flight strips to represent the air situation. Flight strips are ordered and placed specially in front of a controller to give the controller an idea about the relative position of the aircraft. Aircraft position determination and position updates are made by either pilot reports or visual observations. Position estimates are used to determine if there is a potential conflict between aircraft being at the same level.

A flight strip can be in paper or electronic tabular form and contains planned and current flight plan data for a specific flight, Figure 4.14. Depending on the form, paper or electronic, it has to be available on an electronic display or flight progress board to be used by air traffic controllers who provide ATS. There are many styles of progress strip layouts, but each flight strip contains at least:

- Aircraft identification (e.g., aircraft registration or flight number);
- Aircraft type, as the relevant 4-letter ICAO designator (e.g., B744 for a Boeing 747-400);
- Level (assigned altitude);
- Departure and destination;
- At least one time in four figures (other times can be shortened to minutes only).

Other information may be added as required. Many ATSs use several layouts, e.g., one for departing, another for arriving, and a third one for transit overflying aircraft when the unit is the Control Tower or Approach.

Electronic strips have been developed as a response to paper strip limitations (e.g., limitations in information sharing, low transparency due to editing, etc.). Their look and feel are similar to paper strips in terms of layout, colours, and strip management. Electronic strips allow for real-time data updates, streamlining the entire flight-plan process, and enabling data sharing with stakeholders for collaborative decision making around surface movement.

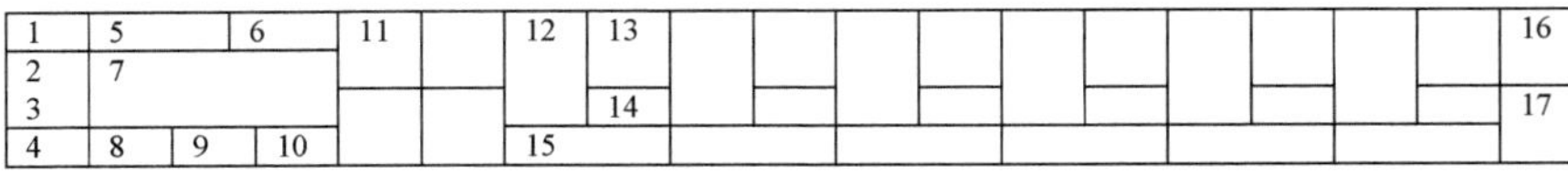

Figure 4.14. Flight progress strips.
Source: Babić and Netjasov 2016

*1. Verification symbol; 2. Transfer of control data and coordination indicators; 3. Time zone; 4. Date; 5. Type of aircraft and requested FL; 6. SSR transponder code; 7. Aircraft identification; 8. Departure airport; 9. True airspeed; 10. Destination airport; 11. Approved FL; 12. Time over previous fix (h); 13. Estimated time over previous fix (min); 14. Actual time over previous fix (min); 15. Previous fix; 16. Sector number; 17. Designated exit corridor.

Radar control system is a method of providing air traffic control services and is applied within controlled airspace for identification, tracking, guidance, and safe separation of aircraft in all phases of flight. This system uses data from radar, Automatic Dependent Surveillance-Broadcast (ADS–B), and Multilateration (MLAT) system for providing the surveillance service (ICAO Doc 4444 2016). When surveillance systems are used (based on radar, ADS-B, or MLAT), the minimum separation prescribed by ICAO Doc 4444 is 5 NM. This minimum may be reduced by the appropriate ATS authority, but not below:

- 3 NM when the surveillance systems' capabilities at a given location permit this;
- 2.5 NM between succeeding aircraft which are established on the same final approach track within 10 NM of the runway threshold. Several additional criteria must be met to apply this separation minimum (described in detail in ICAO Doc 4444 8.7.3.2 b).

Controllers receive data to be used in the radar system from the following data sources:

- Data on aircraft intentions and other data are obtained from the flight plan;
- Data for coordination and transfer of air traffic control is received by telephone or intercom;
- Meteorological data in the form of reports are received from meteorological stations, forecasts are received from meteorological stations at the airport, and information and warnings are received from competent meteorological observation centres;
- Information on the position of aircraft in the airspace is received directly from the surveillance radar indicator at the controller's workplace.

When an aircraft takes off, the aircraft surveillance begins by radars (called Primary Surveillance Radar and Secondary Surveillance Radar), ADS-B and MLAT systems. These systems may be used either alone or in combination to provide air traffic services, including the provision of separation between aircraft.

Primary Surveillance Radar (PSR) uses reflected radio signals and records distance and directs aircraft and displays the given data on the radar screen placed at the workplace of the air traffic controller (ICAO Doc 4444 2016). It is often used in circumstances where other ATS surveillance systems alone would not meet the air traffic services requirements. PSR is the only surveillance sensor used in civil aviation that does not require any on-board equipment to locate aircraft.

Secondary Surveillance Radar (SSR) is a transceiver that emits appropriate query signals, and in response receives data on the altitude and identification code of the aircraft. It uses transmitters/receivers and transponders (ICAO Doc 4444 2016). Secondary radar allows controllers to determine the direction and distance of the aircraft in relation to the radar antenna, the height of the aircraft, then, to identify the aircraft based on the individually assigned aircraft code, as well as to communicate with a particular aircraft. SSR may be used alone and may provide separation between aircraft.

Automatic Dependent Surveillance—Broadcast (ADS-B) is a surveillance technology by which aircraft, aerodrome vehicles, and other objects can automatically transmit and/or receive data such as identification, position, and additional data, as appropriate, in a broadcast mode via a data link (ICAO Doc 4444 2016). It provides real-time precision, shared situational awareness, and advanced applications for pilots and controllers. ADS-B relies on aircraft or airport vehicles broadcasting their identity, position, and other information. ADS-B involves aircraft/airport vehicles using a certified position source to determine their position and broadcasting it in short intervals using a data link in the radio frequency spectrum (SESAR 2020). This signal can be captured for surveillance purposes on the ground (ADS-B Out) or onboard other aircraft to facilitate airborne traffic situational awareness, spacing, separation, and self-separation (ADS-B In). Its implementation in conjunction with existing surveillance systems may contribute to enhancing accuracy, data availability, and reducing frequency load.

Multilateration (MLAT) system is a group of equipment configured to provide position derived from SSR transponder signals (replies or squitters) primarily using time difference of arrival techniques (ICAO Doc 4444 2016). As a navigation tool, multilateration uses a single, mobile receiver to measure the signals transmitted from several sites at fixed, known positions. Multilateration can also be used for both ground (aerodrome) and air traffic surveillance. MLAT sensors (transmitter/receivers) can be sited within the airport boundaries for ground surveillance, at and near an airport for monitoring arrival and departure traffic or over an extended area, where conventional radar is not practical or possible, for monitoring en-route traffic.

4.9 Air traffic controllers' responsibilities

The role of air traffic controllers is to ensure the safe, efficient, and expeditious flow of air traffic through the airspace for which they have responsibility. During their working time, they direct air traffic within the assigned airspace, and control moving aircraft and service vehicles at airports. Certified controllers need to possess specific knowledge, skills, and ability to perform air traffic controller duties, as well as comprehensive knowledge of air traffic control laws, rules, and regulations. Controllers must remain alert and effective throughout their assigned shift which involves besides the regular operational duty, a readiness to cope with unexpected or unforeseen situations, such as the consequences of aircraft emergency declarations, mid-air collision situations, or extreme weather conditions (Eurocontrol Experimental Centre 2004).

Primary responsibilities of air traffic controllers, *inter alia,* are to:

- Compile information about planned and current flights from flight plans, pilot reports, radar, and observations;
- Control flow of air traffic within assigned airspace using radar monitors, radio and other communication equipment and visual references;
- Direct activities of all moving aircraft and service vehicles on or near airport runways;
- Identify aircraft and ensure the inviolability of the assigned airspace;
- Determine mutual positions and ensure the prescribed separation of aircraft;
- Monitor and direct the movement of aircraft, using radar equipment;
- Prevent aircraft collisions (in the air and on the ground);
- Provide pilots with information on weather conditions, weather updates, navigation equipment, condition of manoeuvring areas of the airport;
- Maintain radio and telephone contact with adjacent control towers, terminal control units, and other area control centres to coordinate aircraft movement;
- Alert airport response staff in the event of an aircraft emergency, initiate and coordinate searches for missing aircraft.

There are different types of air traffic controllers, with different skill sets and responsibilities based on their roles, viz.: aerodrome controllers, approach and departure controllers, and area controllers.

Aerodrome Controllers are based in the Control Tower at an airport and are sometimes referred to as "tower controllers". They are responsible for the safety and movement of vehicles, including aircraft, on runways and taxiways. They check flight plans, give pilots clearances to take off and land, and direct the movement of aircraft around the airfield so that aircraft can taxi safely between runways, stands, and other areas they need to travel to.

Approach and Departure Controllers are based at airports and ensure that aircraft travelling within an airport's airspace maintain minimum separation for safety. They give clearances to aircraft approaching the airport to enter controlled airspace and put them into the most efficient position to land. They are also responsible for departing aircraft in their initial phases of flight. Also, they inform pilots about weather conditions and other critical notices.

Area Controllers work at area (en-route) control centres located throughout the country which typically are not located at airports. They monitor aircraft once aircraft leave an airport's airspace. They are responsible for aircraft at higher altitudes i.e. for aircraft in the climb, descent, and en-route phases of the flight. They are sometimes referred to as "en-route controllers". Area controllers issue levels, headings, and speeds to separate aircraft, to provide safe and expeditious routing of traffic within their sectors.

ATC Unit Coordination. The coordination and transfer of control over a flight between successive ATC units and control sectors are done by a dialogue comprising the following stages (ICAO Doc 4444 2016):

- Notification of the flight to prepare for coordination, as necessary;
- Coordination of conditions of transfer of control by the transferring ATC unit;
- Coordination, if necessary, and acceptance of conditions of transfer of control by the accepting ATC unit;
- Transfer of control to the accepting ATC unit or control sector.

Coordination procedures between ATC units are specified in letters of agreement and local instructions. The responsibility for the control of an aircraft is transferred from an ATC unit to the next unit at the time of crossing the common control area boundary as determined by the unit having control of the aircraft or at such other point or time as has been agreed between the two units. If the flying time from the departure aerodrome of an aircraft to the boundary of an adjacent control area is less than the specified minimum required to permit the transmission of the necessary flight plan and control information to the accepting ATC unit after take-off and allow adequate time for the reception, analysis, and coordination, the transferring ATC unit shall, before departure, forward that information to the accepting ATC unit together with a request for approval (ICAO Doc 4444 2016).

There are two types of coordination, verbal and non-verbal. Verbal coordination is usually done using telephone lines, but in some cases, a dedicated hotline connection can be established (usually for coordination between tower and approach). Verbal coordination is used in complex situations or where electronic coordination have not been implemented. Typical verbal coordination lasts about 30 seconds, but can take more than a minute for complex situations (e.g., involving multiple aircraft or in case of an emergency or unusual situation) or if additional information is required (e.g., full flight plan messages, weather reports, etc.). Electronic coordinations are generally used in routine and less complex situations. Various electronic coordinations may be available, including release procedure, coordination for a direct route or transferring an aircraft on a heading, the ability to counter-propose (a heading, level, fix, etc.), the ability to highlight a particular aircraft to support the phone coordination and many others. They require less time to perform (usually a few seconds) and generally lower the controller's workload.

The most common coordination and coordination-related procedures include (ICAO Doc 4444 2016):

- *Exchange of flight information* (e.g., basic and extended flight data, entry clearances, etc.);
- *Approval requests*. An approval request may be a part of the initial flight information exchange;
- *Transfer of control and communication*. The transfer of control is usually effected at the moment an aircraft passes the boundary between the two ATS units. Transfer of communication, however, usually takes place earlier (1–2 minutes in most cases). The receiving controller is not supposed to modify the ATC clearance of the aircraft prior to the boundary crossing unless the release procedure has been coordinated (see next point);
- *Release*. This procedure is a part of the transfer of control and communication. The transferring controller delegates control of an aircraft to the receiving controller whilst the aircraft is still within the transferring controller's airspace. The responsibility for separation provision lies with the transferring controller. The release may be

restricted at the transferring controller's discretion, e.g., the aircraft may only be released for the descent to a specified level;

- *Notification.* These procedures include informing the neighbouring unit of certain situations (e.g., radar or another equipment failure, aircraft emergency, aircraft not contacting the receiving unit, etc.);
- *Rejection and counter-proposal.* If a controller cannot accept an aircraft under the conditions specified, they may reject the coordination and offer terms under which they would.

4.10 Divisions of air traffic services

Each IFR flight is fully controlled by the air traffic control service during all phases of the flight. To accomplish all the objectives in terms of the safe, efficient, and expeditious flow of air traffic, air traffic control service is divided into three divisions (ICAO Annex 11 2018):

- *Aerodrome control service* that provides air traffic control services to traffic on airport manoeuvring areas, as well as to all aircraft flying in the vicinity of the airport (in controlled areas (CTR)). It is based in the airport control tower. The aerodrome control is responsible for preventing collisions between aircraft flying in the aerodrome circuit, aircraft taking off and landing, and aircraft, vehicles, and obstructions in the manoeuvring area;
- *Approach control service* that provides air traffic control services at the approaches or departures of one or more airports, if they are in the immediate vicinity, i.e., in the final-controlled areas (TMA) and the Control Zone (CTR). Typically, it is based in the airport control tower or area control centre and it is responsible for controlled flights of aircraft arriving at one or more airports under the TMA or departing from them. This unit can be established as a separate unit or by an aerodrome control tower or area control centre, when it is necessary to combine the functions of the approach control service under the responsibility of a single unit;
- *Area control service* that provides air traffic control services for en-route controlled flights in control areas, i.e., in the areas of air traffic control for which they are responsible (CTA/UTA)—called a sector. En-route controllers direct and separate aircraft flying within their sector. Area control service is based in the area control centre. The main responsibility of this unit is to assign cruising levels, deal with crossing traffic, and maintain separation during the final climb and initial descent.

4.11 Routes established for the use of aircraft

An aerodrome controller notifies pilots when it is safe to push the aircraft out of the gate at the airport, issues instructions for using the runway and monitors the aircraft's movement during their taxiing phase. If the runway is available, the aerodrome controller issues a departure clearance and grants permission to enter the runway and depart. After the aircraft takes off, the aerodrome controller will assign the aircraft a frequency change to the departure controller and the aircraft continues to climb to the assigned altitude. Departure controllers take responsibility for the aircraft through their airspace while safely separating the aircraft from other arriving, departing, and transitioning aircraft. This phase ends by entering the cruise phase when controllers in the area control centre take over responsibility for aircraft. They coordinate with pilots on weather conditions and issue instructions on speed, route, and altitude to ensure positive separation from other aircraft operating under IFR. When the aircraft moves into a new sector, the next controller takes over. At this stage, the movement of the aircraft can be monitored using the signals of the navigation devices such as VHF Omnidirectional Radio Range (VOR) and Distance Measuring Equipment (DME).

VOR is the standard air navigational system for aircraft that emits specific signals, the so-called radials, enabling them to find their position and stay on course. Each radial has a numerical value, which determines the angle position of the aircraft relative to the VOR. A VOR ground station does not measure the distance of the aircraft from this device. Typically, to measure the distance as well, a DME navigation device is installed in addition to the VOR device. With these two devices, used as route fixes within high or low airspace, each aircraft can perform its flight as planned, along the entire route. It is also possible to place PSR/SSR radars along the route (usually at the top of a mountain along the route) to allow flight surveillance during a cruise from the area control room.

Long range routes are designed to be the most efficient way for aircraft to get from point A to point B, on the other side of the world. Although it looks like they are flying a very long route, aircraft are moving on the shortest route connecting two points on the earth, following a route that coincides with a large circle of the sphere (orthodrome[3]). It would be easiest for pilots to always follow the route that coincides with the large circle, however, in reality, this is not always the case.

On the part of the route that is above the mainland, pilots follow navigation devices that are on the ground (VOR/DME) or if they have the appropriate equipment, they use the Global Positioning System

[3] Orthodrome is the intersection of the sphere and a plane that passes through the center point of the sphere.

(GPS) system. When moving along a given route, the aircraft flies over the navigation devices on the ground and thus allows the controller to determine its current position. The part of the route that is above the ocean cannot be marked with appropriate navigation devices, nor surveillance radars can be set up, so flight tracking is done differently. On its flight across the ocean, the aircraft follows a route that, along the entire route across the ocean, is predefined by points with known coordinates (longitude and latitude). The aircraft enters and exits the route above the ocean at precisely defined points on both sides of the ocean so that controllers can estimate its approximate position during a flight on this part of the route.

Instrument landing system (ILS) is designed to provide the pilot with an approach path that is perfectly aligned with the runway centreline (Nolan 2011). An ILS provides both lateral and vertical guidance during the approach to land.

The ILS system is equipped with three different types of transmitters: the localizer, the glide slope, and two or three marker beacons (Nolan 2011). The localizer provides azimuth guidance. It is normally located at the end of the runway and provides the pilot with lateral course guidance information only. The equipment in the aircraft provides information to the pilot showing the aircraft's displacement from the runway centreline. The glideslope defines the correct vertical descent profile. The equipment in the aircraft indicates the displacement of the aircraft above or below the glideslope. Marker beacons and high intensity runways lights may also be provided as aids to the use of an ILS. They are normally located on the ILS approach track at notified distances from touch-down.

When the runway visual range (RVR) is above the specified minimum and the approach is flown, the pilot follows the ILS guidance until the decision height (DH) is reached. There are defined categories of ILS approaches that are allowed for suitably qualified pilots flying suitably equipped aircraft to suitably equipped runways using appropriately qualified ILS systems (Nolan 2011):

- Category I permits a DH of not lower than 200 ft and an RVR not less than 550 m;
- Category II permits a DH of not lower than 100 ft and an RVR not less than 300 m;
- Category IIIA permits a DH below 100 ft and an RVR not below 200 m;
- Category IIIB permits a DH below 50 ft and an RVR not less than 50 m;
- Category IIIC is a full auto-land with roll out guidance along the runway centreline and no DH or RVR limitations apply.

Aviation is increasingly moving to the new navigation systems technology, i.e., GPS-based navigation, which improves safety and flight efficiency by enabling aircraft to fly direct from departure to destination using the most fuel-efficient routes. GPS is a space-based positioning, velocity, and time system, which enables three-dimensional position determination for all phases of flight from departure, en-route, and arrival, to airport surface navigation. Greater use of GPS navigation system allows aircraft to fly user-preferred routes from waypoint to waypoint, where waypoints do not depend on ground infrastructure. This feature is of particular importance in areas that lack suitable ground based navigation aids or surveillance equipment.

GPS navigation system allows the establishment of new and more efficient air routes. Moreover, satellite navigation permits accurate aircraft position determination anywhere on or near the surface of the earth and, thus, allows to safely reduce the separation minima between aircraft in the airspace, resulting in increased capacity and capabilities.

Free route airspace (FRA) is a specified volume of airspace in which users can freely plan a route between defined entry and exit points. Pilots can plan their direct route via intermediate waypoints, without reference to the air traffic service (ATS) route network. Inside this airspace, flights remain subject to air traffic control (Eurocontrol 2015). FRA was initiated in Europe in 2008 and included in the Flight Efficiency Plan developed cooperatively by International Air Transport Association (IATA), Civil Air Navigation Services Organisation (CANSO), and Eurocontrol (IATA/CANSO/Eurocontrol 2008).

In the FRA concept, all aircraft are allowed to fly their optimal route and route selection is subject to only a few limitations (e.g., fixed entry and exit points and the need to avoid danger areas or TSAs). In most cases, the straight line between an entry point and an exit point will be chosen. If for some reason this is not appropriate, additional turning points can be specified. These can be navigational aids, published navigational points, or points with specified coordinates. Therefore, the task of traffic separation is moved from air traffic control to the cockpit (airborne separation).

Furthermore, FRA can offer several benefits for airspace users, such as:

- Lower operating expenses—a decrease in flight times and fuel consumption as aircraft follow more optimal operating routes, i.e., are more aligned with user preferences;
- Increased capacity—an increase in available airspace capacity, which further leads to improved traffic predictability due to more stable trajectories;
- Lower environmental impact—a reduction in aviation emissions as a consequence of reduced flight times.

FRA concept will be implemented as Class C airspace, with certain agreed exemptions. Airspace reservations will remain, and as all airspace users will have equal access to FRA, harmonised application of the FUA concept and Civil/Military Coordination are taken into account to ensure harmonised procedures and service provision for the benefit of all airspace users.

By the end of 2017, 51 area control centres either fully or partially implemented the FRA (Bentrup and Hoffman 2016). Most of these free route airspaces are either time-limited (e.g., at night) or structurally limited (restricting entry or exit points for certain traffic flows, applicable within control areas or upper airspace only). The FRA with time limitation is usually a transitional step that facilitates the early implementation and minimises safety risks. The structurally limited FRA is more common in complex airspaces where full implementation can hurt safety and capacity.

References

Babić, O. and Netjasov, F. 2016. Air Traffic Control, University of Belgrade, Faculty of Transport and Traffic Engineering.

Bentrup, L. and Hoffman, M. 2016. Free Routing Airspace in Europe: Implementation Concepts and Benefits for Airspace Users. 7th International Conference on Research in Air Transport, ICRAT.

Eurocontrol. 2018. Long-Term Forecast: IFR Flight Movements 2010–2030. Available at: http://www.eurocontrol.int/statfor/gallery/content/public/forecasts/Doc415-LTF10-Report-Vol1.pdf. Accessed: 27.02.2018.

Eurocontrol. 2000. Eight States Free Route Airspace Project – Free Route Airspace Concept. Available at: https://www.eurocontrol.int/sites/default/files/library/007_Eight-states_Free_Route_Airspace_Project.pdf. Accessed: 29.09.2020.

Eurocontrol Experimental Centre (EEC). 2003. Pessimistic sector capacity estimation. EEC Note Number 21/03.

Eurocontrol Experimental Centre. 2004. Critical Incident Stress Management in Air Traffic Control, EEC Note No. 15/04, Project SAF–D–E3-000.

FABEC. 2014. ATM in Europe. It's all about performance. Available at: https://www.fabec.eu/images/user-pics/pdf-downloads/FABEC_All_about_Performance_1.pdf. Accessed: 14.08.2020.

IATA/CANSO/EUROCONTROL. 2008. Flight Efficiency Plan. Fuel and emissions savings. Available at: https://www.eurocontrol.int/sites/default/files/2019-04/airspace-flight-efficiency-plan-aug2008.pdf. Accessed: 08.09.2020.

ICAO. 1984. Air Traffic Services Planning Manual, Doc 9426, First Edition.

ICAO. 2005. Annex 2 to the Convention on International Civil Aviation – Rules of the Air. Tenth edition.

ICAO. 2011. Civil/Military Cooperation in Air Traffic Management Cir330 AN/189.

ICAO. 2012. Manual on a 300 m (1 000 ft) Vertical Separation Minimum Between FL 290 and FL 410 Inclusive, Doc 9574-AN/934, Third Edition.

ICAO. 2016. Procedures for Air Navigation Services – Air Traffic Management (PANS-ATM), Doc 4444-ATM/501, Sixteenth Edition.

ICAO. 2018. Annex 11 to the Convention on International Civil Aviation - Air Traffic Control Service, Flight Information Service, Alerting Service. Fifteenth edition.

ICAO. 2018. Procedures for Air Navigation Services – Aircraft Operations, Volume I – Flight Procedures, Doc 8168, Sixth Edition.

Nolan, M.S. 2011. Fundamentals of Air Traffic Control, Clifton Park, N.Y.: Delmar Cengage Learning, 5th Ed.

Regulation (EC) No. 549/2004 of 10th March. 2004: Single European Sky; Framework Regulation.

SESAR. 2020. Automatic Dependent Surveillance – Broadcast. https://ads-b-europe.eu/, Accessed: 24.09.2020.

Chapter 5

Airlines

An airline is a company that provides air transport service for passengers and goods. Airline possesses its fleet, routes, operating personnel, and management to perform the offered services.

Airlines can vary in fleet size, service offered, network structure, route lengths, etc. They can have a small number of aircraft in their fleet, but they can also be large international airlines that offer full-service and have a fleet of several hundred aircraft. The aircraft in their fleets can be different in size/capacity and range. They can perform intercontinental, continental, regional or domestic flights. Some airlines specialize only in the transport of cargo, some only in the transport of passengers, and there are those that transport both passengers and cargo. It is also possible to distinguish between airlines that perform scheduled traffic and those that offer charter transport service. So, there are various criteria on the basis of which the division of airlines can be made.

There are essentially five types of airlines, as follows:

- Full service carrier (FSC)—offers high levels of in-flight service, connectivity and tariffs, attracts various passenger segments, operates on network of short-, mid- and long-haul routes and poses the fleet of variety of aircraft types;
- Regional carrier—connects smaller airports with large ones, has aircraft with smaller capacity in its fleet compared to FSC;
- Low cost carrier (LCC)—an airline with cost-minimization strategy in order to provide low operating costs and prices;
- Charter carrier—an airline that performs non-scheduled flights mostly to holiday destinations;
- Cargo carrier—an airline specialized in transport of goods by air.

It should be noted that this is a useful guide for airline business model differentiation, but these categories are not absolute and some overlap exists between them. For example, FSC can also operate charter flights, as well as transport cargo. Also, FSC operates in regional short-haul markets.

Airlines are most often ranked on the basis of realized passenger kilometres, which is actually a measure of realized passenger traffic. This measure is calculated as the sum of the products of the number of passengers who paid for the air ticket on each flight and the distance the aircraft flew. In the literature, this measure is called revenue passenger-kilometre (RPK) or revenue passenger-mile (RPM), depending on the distance unit used (Shaw 2020). In the cargo industry, the realized traffic is measured by revenue tonne-kilometre (RTK) that is defined as a tonne of payload (cargo) flown over one kilometre (Shaw 2020).

The indicator of the airline product is the available seat kilometres (ASKs). It is calculated as the sum of the products of the number of seats on each flight offered to passengers and the distances flown by the aircraft (Holloway 2008). This measure is actually a measure of available capacity for one airline. The utilization of capacity is termed 'load factor'. The passenger load factor for one flight represents the ratio of the number of passengers on the flight and the number of available seats on the aircraft used. The passenger load factor for the airline network, in a certain period of time (day, month, and year) represents the percentage share of total realized passenger kilometres (RPK) in the total available seat-kilometres (ASK) (Holloway 2008).

5.1 Full service carriers

Full service carriers (FSCs) are typical former national or major airlines that have large networks with intensive traffic on the routes and provide full service that includes: serving major airports, offer different cabin classes (economy or business class travel and on some flights premium economy and first class.); in-flight service; in-flight entertainment (movies, music, information on display during a long-haul flight); sale of perfumes, drinks, cigarettes and other things on the plane during the flight, rewarding frequent travellers, special service for business and first class passengers or passengers-members of the loyalty program (frequent flyer program), airport lounge usage while waiting for the flight after check-in or when transferring. In Europe, FSCs are most often former national airlines that have been privatized to some extent over the years.

FSCs operate hub-and-spoke networks which arose as a result of the mix of a hub-and-spoke and point-to-point networks, previously operated in the regulatory regime. This type of network structure allows an airline to benefit from a number of cost and demand advantages. Hub-and-spoke networks require the concentration of traffic on one or two central hub airports where passengers can change aircraft on their way to their final destinations (Berry et al. 1996). The FSC uses the hub-and-spoke route structure in order to feed and manage the whole network to optimise seat load factors (Doganis 2010).

FSCs are the main carriers for the scheduled and charter traffic (passenger and freight) in many countries and they are the traffic rights holders from international bilateral agreements. These airlines in Europe perform almost all intercontinental traffic and most of the medium-haul traffic. They also perform part of the traffic on short-haul routes, especially on those routes with strong traffic flows. Most of these airlines are cooperating with one or more regional airlines through ownership and commercial contracts (Krstić and Stepanović 2004).

The fleets used by FSCs are typically composed of the variety of aircraft due to the diversity of routes on which they perform their service. Usually, these type of airlines own the companies that provide other services in air transport, such as ground handling agencies at airports, aircraft maintenance, catering, etc. The FSCs adopted frequent-flyer programs (FFP) designed to induce and maintain customer loyalty, especially among business travellers (Wensveen 2015). Moreover, for the ticket reservation purposes they developed sophisticated reservations systems that interfaced with a global distribution system (GDS) which supports travel agencies and other distribution channels in making reservations for most major airlines. Most of the FSCs are within one of the world's airline alliances (SkyTeam, oneworld, Star Alliance).

In response to the emergence of low-cost airlines, FSCs started to create subsidiaries that offer a transport service with reduced price and quality of service (their own low-cost airlines). This is one of their survival strategies, driving them to reduce operating costs, withdrawing from unprofitable routes, and offering low prices on competitive routes. However, it should be noted that even before the emergence of low-cost airlines, there was a constant trend of reducing the prices through a range of tariffs valid under certain travel conditions (group fare, minimum/maximum stay, weekend coverage, restrictions on changing reservations, etc.). Therefore, lower price of air travel is a consequence of evolution in technical development, improvement of organization and improved efficiency of airline operations.

The FSC business model is followed by airlines such as Lufthansa, Air France, British Airways, Turkish Airlines, United, Delta, American Airlines, Singapore Airlines, Qantas, Japan Airlines, Air Canada etc. There are many differences between the business strategies adopted by these airlines, but the key aspects of their business models listed in this chapter are those they generally share in common.

5.2 Regional airlines

One of the fastest growing segments in the airline industry in recent years is regional air service. In the following chapters, the concept and the main characteristics of regional traffic will be explained. The regional airlines and the phases of their development will also be described. And finally, associations of regional carriers, with reference to the differences between the European market and markets of the United States will be discussed.

5.2.1 The concept and basic characteristics of regional traffic

It could be said that the regional traffic is the traffic performed by turboprop (TP) or small jet (regional jet—RJ) aircraft with the capacity of up to 100 passenger seats and/or traffic performed at regional airports which are situated far away from large cities. This could be considered as one of the first definitions in its original form. However, the market conditions have been changing over years, therefore the previously stated definition has also evolved. New trends in regional traffic indicate that the size of aircraft is increasing, which means that regional traffic is also performed by aircraft with a larger capacity (sometimes more than 100 passenger seats). Another change concerns airports. Initially, regional traffic was performed only at regional airports, but now there is a tendency to connect local and larger airports, the so-called hubs. Therefore, it could be said that the regional traffic provides feeder service for traditional (full) service airlines and play the main role in connecting small regional airports and small communities with the large hubs and larger cities, offering a large variety of connections to passengers.

Bearing in mind that the hubs are congested and have a problem with the available capacity, the importance and use of regional airports is increasing. Furthermore, small, regional aircraft (RJs and TPs) are very suitable for operations at smaller, regional airports. In order to differentiate regional air traffic from others, some basic characteristics of regional airlines are introduced, and they are:

- the usage of regional aircraft, either RJs or TPs, with a capacity of up to 100 passenger seats;
- operating on short-haul and medium routes;
- traffic is performed on routes of low and medium traffic density (10000–100000 passengers per year);
- high frequency (at least 2 flights per day, e.g., early morning and late evening flights that allow departure and return on the same day, enabling users to reduce travel time and travel costs);
- regional airlines are not competitors to traditional airlines (traditional, unlike regional airlines, have large aircraft with a capacity of over

100 passenger seats in their fleet, and their flight schedules do not overlap);

- route networks of regional airlines are complementary to the networks of traditional airlines;
  - o they cover routes which are not attractive to traditional airlines due to inappropriate aircraft in their fleet;
  - o they satisfy demand during the working week on routes of low and medium traffic density;
- most traditional airlines have their own regional subsidiaries or appropriate cooperation agreements with some regional airlines (connecting smaller airports (areas with low passenger demand) with large hubs, providing feeder service); and
- they are primarily focused on business travellers.

5.2.2 The development of regional airlines

The development of regional airlines could be divided into several phases.

The first, initial phase of the development is related to the early beginning of an airline which provides regional services at the airline market. This phase is characterized by offering flights with a small aircraft capacity on short destinations, usually within the state borders. Large airports and hubs are avoided, and only local, regional centres are connected. Airlines are independent (the ownership is usually private).

The second phase is the development phase. Airlines are increasing the frequency of flights, changing the route network, changing the fleet structure, looking for a suitable place in the airline market. The capacity of aircraft in the fleet is increasing, as well as distance between origin and destination airport.

The third phase of development of regional airlines is the phase characterized by changing the ownership structure of airlines. When independent regional airlines affirm themselves on the airline market so that they could in some way endanger the interests of other airlines (primarily traditional ones), the interest of traditional airlines arises, which results in mutual cooperation. Cooperation occurs for two reasons:

1. Traditional airlines have large capacity aircraft in their fleet, which disable them to operate weaker routes; the capacity would be underused, i.e., passenger load factor would be law, and traditional airlines must reduce the frequency in order to adjust to demand; reducing the frequency would further lead to a decrease in the number of passengers (passengers, especially business ones, are time-sensitive and the flights frequency and take-off times are decisive factors for

them); the traditional airline perceive a need for cooperation with regional airlines that would fly for them on short, weaker routes.

2. Another reason is that a regional airline could establish cooperation with the other airline and thus feed other hubs, which would not suit a traditional airline, because passenger flows would be redirected to other centres.

Therefore, the traditional airlines outsource their services/flights to regional airlines. In this way, the regional airlines started to work for the traditional airlines, begin to connect local centres and hubs and gradually loses its independence or become fully owned by the traditional airlines (some examples: Tyrolean Airways which became fully owned by Austrian Airlines, Lufthansa CityLine which is owned by Lufthansa).

The fourth phase of development involves the formation of airline groups, which was initially characteristic of the US market (American Eagle—American Eagle Airlines and Executive Airlines in the past, today Envoy Air—Envoy Air and Executive Airlines, SkyWest Airlines—SkyWest Airlines and Atlantic Southeast Airlines once, today SkyWest Airlines—SkyWest Airlines and Expressjet Airlines). For a while this happens in EU market as well; Lufthansa Regional consisted of Air Dolomites, Augsburg Airways, CityLine, Contact Air and Eurowings, and after that reduced the number of airlines to Air Dolomites and CityLine, and finally the name Lufthansa Regional stopped to appear since 2018. However, regional airlines can retain its independence and create the partnerships with different airlines (for example, SkyWest works with large airlines such as Delta, United, and Alaska).

It is worth noting some remarks related to differences and similarities between regional and low-cost airlines on the EU and US markets considering provision of air services on thin routes. These thin routes are attractive for both LCCs and regional airlines. However, some differences exist. It is shown by Fageda and Flores-Fillol (2012), that the main difference is related to route choices. Precisely, in the US' business model of LCCs is close to traditional airlines due to the fact that both types of airline prefer to operate on high-density routes. On the other hand, thin routes are operated mostly by regional airlines which use RJs. Considering the EU market, it can be noted that thin routes are mainly operated by LCCs. Regional services provided by RJs in the USA enable airlines to offer higher frequency, but the higher fares as well, and it is more attractive for business travellers. On the contrary, low-cost services operated by mainline jets enable airlines to offer lower fares, but the flight frequency is reduced and it appeals to leisure travellers (Fageda and Flores-Fillol 2012).

5.2.3 Association of regional airlines

Regional airlines have their associations. The well-known are European Regions Airline Association (ERAA) and Regional Airline Association (RAA).

ERAA is the European body for regional traffic. It was founded in 1980, and at the beginning it consists of five members. Today, ERAA represents more than 200 members, including more than 60 airlines, and around 150 associated companies which provide different services (airports, manufacturers, suppliers, etc.). According to ERAA (2021) the airlines members of the association have a fleet which consists of 730 passenger aircraft and 105 cargo aircraft. Regarding the share of jet and prop aircraft, it can be seen that 57% are jet aircraft, while 43% are turbo-prop. Airlines operate on 2742 routes, offering 1.12 million of flights per year. The average distance length is approximately 650 km, while average seating capacity is 80 seats.

RAA is the US regional association and it was founded in 1975 as the Commuter Airline Association of America. This association represents US regional airlines and other suppliers. As regional services (or commuter) are very common in the USA and more developed than in Europe, it is not surprising that more than 40% of all scheduled passengers fly by RAA members. According to RAA (2020) there are 66 US regional certified carriers, which completed 3.81 million departures (40% of all scheduled passenger departures in the USA) carrying 165.17 million passengers in 2019, with average seating capacity of 64 seats and distance length of nearly 800 km on average. Airline members operate daily more than 10000 flights on average, with a fleet comprising 1972 aircraft. The volume of regional air traffic in the USA is large, and approximately 66% of airports are dedicated to regional airlines only.

The number of regional passengers is constantly growing from 1980, but reaching of saturation level could be noticed from 2007 (Figure 5.1).

The leading regional airlines in the USA according to number of passengers (RAA data) are presented in Table 5.1.

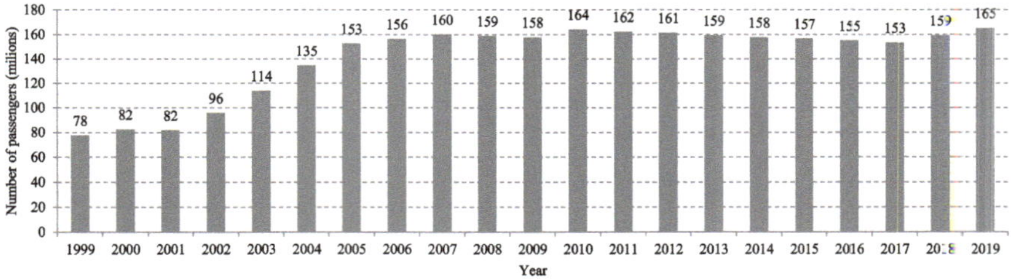

Figure 5.1. Number of regional passengers in the USA (*data source*: RAA 2021).

Table 5.1. Leading RAA's airlines in 2019.

Rank	Airline	Number of passengers (millions)
1	SkyWest Airlines	42.46
2	Republic Airlines	18.88
3	Endeavor Air Inc.	15.15
4	PSA Airlines	15.12
5	Envoy Air	15.08
6	Mesa Airlines Inc.	15.06
7	Horizon Air	7.85
8	Compass Airlines	6.35
9	ExpressJet Airlines Inc.	6.05
10	Piedmont Airlines	5.04

Data source: RAA 2020, Annual Report

5.3 Low-cost carriers

Low-cost carrier[1] (LCC) is an airline that offers its passengers low fares by eliminating many traditional travel services which do not provide added value (Schlumberger and Weisskopf 2014). Although they are recognizable by low fares, this airline category is defined by a focus on generating lower operating costs in comparison to other, competing airlines. These cost savings are transferred to consumers in the form of lower fares.

The development of the low-cost business model was closely linked to the deregulation of the air transport market in the United States (1978), when new air carriers were allowed to enter the market freely (Fischer 2018). After this idea was successfully implemented and proven to be sustainable over time in the United States, the business model itself, along with deregulation, has spread first to Europe and then to the rest of the world.[2] The market share of low-cost airlines varies across continents and ranges from about 10% in the Asian, 20% in the European and about 30% in the US market (ICAO 2019).

1 Other names value-based carrier, no-frills carrier or low fare carrier.

2 Southwest Airlines is considered as original low-cost airline that developed the low-cost business model. The strategy of this airline has always been to offer a service on the market, which will be competitive not only with other airlines, but also with other types of transportation. Southwest's business model is based on extremely efficient operations, low-cost pricing and focus on customer experience. Beside Southwest, there are other successful low-cost airlines in the US today, such as Frontier and JetBlue. In Europe, the first established LCC was Ryanair from Ireland (1986). However, the biggest wave of low-cost airlines occurred after 1997, when the deregulation and liberalization of the market in the European Union was completed. The LCCs that appeared in this period were Go, Buzz, Easyjet, Wizzair, etc. After Europe, low-cost airlines began to appear in other parts of the world, the most successful among others are Air Asia in Malaysia and Virgin Blue in Australia.

A successful low-cost airline business model can be defined by three key elements: simple product (no in-flight meals, no free luggage, inability to reserve seats, etc.), low operating costs (high employee productivity, densely spaced cabin seats, use secondary airports, homogeneous fleet, etc.) and market positioning (direct, non-stop flights, high frequencies on routes, high-speed service at airports, etc.). The following sections examine all these characteristics in terms of service offering, network structure, fleet and labour utilization which have enabled LCC to achieve cost savings and therefore offer lower fares (Schlumberger and Weisskopf 2014).

Service offering. Low-cost carriers offer a simple product through simple operations. The LCC's offer implies low ticket prices on the market, which is achieved by reducing the costs, both operational and non-operational. One of the ways to achieve this goal is to eliminate all those services that are a great financial burden for the airline, on the one hand, and do not significantly affect the airline choice by passengers, on the other hand. Accordingly, LCCs have eliminated the service of serving food and drinks during the flight. Drinks and food on their flights are only available if charged, but in that case this service should be seen as an additional source of revenue.

The reduced service of LCCs does not only mean the abolition of food and beverages during the flight, but also all other services which increase operating costs and that are not part of the standard transport service (e.g., in-flight entertainment programs, loyalty programs, airport lounges, etc.) (ELFAA 2004). Accordingly, their aircraft are configured with a single passenger class (economy class). Abandonment of business and other upper classes service, which typically have larger seats and require more staff, has allowed LCCs to increase the number of seats in the passenger compartment and to reduce the number of employees per passenger (ELFAA 2004). Also, the seat configuration in the passenger cabin allows them to set the number of seats to maximum (costs per seat are reduced) by using the narrower seats than those of traditional airlines, with less seat pitch and less legroom.

Many LCCs do not offer the option of choosing the seat on the flight (Schlumberger and Weisskopf 2014). They operate on a first-come, first-served basis. This allows them to accelerate the process at the check-in counters, as well as process during the boarding passengers on an aircraft, and thus reducing the time spent at the airport. Moreover, they were able to charge for this extra service and to increase the revenue.

One of the newer measures of LCCs, which was also introduced in order to further reduce costs, is to limit the amount of baggage (both in the number of bags and in kilograms) that each passenger can take with them on the flight (Vasigh et al. 2018). Any excess of the allowed amount is paid extra. This measure is in accordance with their basic principle that they want to offer a simple product on the market. The introduction of these additional measures indicates that by buying a ticket from LCC, the passenger pays only the seat in the aircraft and basic transport service.

Airline management. LCCs employ less people due to their simplicity of operations. Bearing in mind that labour costs are the largest part of the total costs of airlines, for LCCs, one of the most important tasks is to keep their labour costs under control, as well as to increase their productivity (Schlumberger and Weisskopf 2014). Their practice is to outsource ground-handling, maintenance and other services that need great investment, but do not belong to core operations. In comparison to LCCs, traditional airlines have up to twice the employees per aircraft, Table 5.2.

LCCs are focused on increasing labour productivity that is often measured by available seat kilometres/miles (ASK/ASM) per employee. LCCs pay their employees less per hour than other airlines. However, due to the high productivity of their employees, they are paid high salaries at the end of the month, which in some cases are equal to the salaries of employees in traditional airlines. For example, the salaries of pilots at

Table 5.2. Fleet size and employee number of selected traditional and low cost airlines.

Airline	Country	Fleet	Employees	Employees/ Aircraft
Lufthansa	Germany	360	135,534	376
China Southern Airlines	China	603	100,831	167
Air France–KLM	France/Netherlands	543	81,527	150
All Nippon Airways	Japan	241	33,719	140
American Airlines Group	United States	957	128,900	135
China Eastern Airlines	China	607	77,005	127
United Airlines	United States	777	92,000	118
IAG	Spain/UK/Ireland	570	63,531	111
Delta Air Lines	United States	880	89,000	101
JetBlue	United States	269	21,569	80
Southwest Airlines	United States	753	58,800	78
Frontier Airlines	United States	100	5,150	52
Wizzair	Hungary	135	5,300	39
Ryanair	Ireland	467	17,500	37

Data sources: annual airline reports and sites for 2019.

the Southwest airline are among the highest in the airline industry, but they also have high average flight hours per month, as well as additional responsibilities to help cabin crew clean aircraft and carry luggage (Vasigh et al. 2018).

Airline marketing – Ticket Distribution. Ticket distribution is another segment in operations that provides significant cost savings to LCCs. In order to reduce the costs of selling tickets, they avoid all types of cooperation with travel agencies, tour operators, as well as any other type of intermediary that involves the payment of a commission (global distribution systems—GDS) (Schlumberger and Weisskopf 2014). With the advent of the internet, all LCCs have switched to electronic ticket sales as well as direct ticket sales (through their websites and call centres) to passengers through their sites (ELFAA 2004). In this way, LCC do not pay high charges to agents and do not have the cost of printing tickets (tickets are electronic). Another reason for not using the GDS service is that LCCs have network strategies that do not provide connecting services. Thus, for LCCs this GDS service is useless and very expensive.

Although direct sales are the key channel for most LCCs, some of the major LCCs have begun to use GDS services (Southwest, AirAsia, Vueling, Norwegian or Spirit Air). Of course, this forced the airlines to raise the price of these tickets booked through GDSs.

Airline marketing – Pricing. LCCs generally follow the "simple pricing structure" strategy. This strategy involves fewer fare levels, less restrictions in comparison to traditional airlines, and a convergence of the highest and lowest fares offered. Tickets purchased at low prices are not limited by any additional restrictions, except that they cannot be returned (in case that the passenger cannot make the trip) and that the airline is not obliged to connect the flights (in case of a journey that includes more LCC flights). In some cases, the LCCs put the restriction for tickets that is related to the time of booking, with fares tending to increase closer to the departure date (ELFAA 2004).

LCCs have removed the fare restriction that passengers need to purchase a return ticket in order to get lower prices. They also removed those conditions related to the minimum and maximum stay requirements which were in addition to low priced one way tickets of full service airlines.

Fleet. One of the major characteristics of the LCC business model is the use of a homogenous fleet, i.e., fleet consists of one type of aircraft. There are number of benefits for the airline if having a homogenous fleet, such as reduced cost when purchasing an aircraft, reduced maintenance costs when purchasing spare parts (discount that airline receives when purchasing large quantities of the same parts), reduced training costs for mechanic engineers, flight and cabin crew and increased flexibility in planning operations (e.g., fleet assignment, aircraft rotations and crew planning and scheduling) (Vasigh et al. 2018). In case of disturbances of the daily operational flight schedule (e.g., aircraft failure), the solution of the problem is easier to find, because any aircraft from the fleet can replace that aircraft. The same applies to disturbances caused by the absence of flight/cabin crew. Since the airline must provide a reserve crew for each type of aircraft in the fleet, having a homogeneous fleet also reduces the number of required reserve pilots and cabin staff (smaller number of employees) (Schlumberger and Weisskopf 2014).

Among LCCs, the most commonly used aircraft are those from the Boeing 737 new generation series and the Airbus 320 series. These two types of aircraft have proven to be the best when it comes to their flexibility. They can be used for both short-haul and medium-haul routes, and they are also easily adapted to different demands (the configuration of the seats in the passenger cabin can be changed in the range from 120 to 200 seats). Moreover, in order to fit into the LCC business model an aircraft has to be adaptable in terms of short turnaround time (ELFAA 2004). All LCCs try to prepare their aircraft as quickly as possible for the next flight and look for ways to speed up the process. Another equally important aircraft characteristics for LCCs are reliability (to keep maintenance costs to a minimum and to enable a high frequency service) and efficiency (to have low operating costs) (Schlumberger and Weisskopf 2014).

Although the single aircraft fleet has been one of the recognizable features of the LCC business model, this is not accepted by all LCCs. For example, Easyjet and Jetblue have their fleets encompassing two aircraft types because they found that having multiple aircraft sizes in the fleet gives a greater level of flexibility to cope with fluctuating demand.

Network structure and scheduling. LCCs are oriented towards scheduled, short-haul and medium-haul flights. Their network structure is point-to-point, serving only direct flights between city-pairs. Most LCCs do not offer connecting flights, so they are able to continually use airport facilities during the day and utilize employee services more evenly. Moreover, without connecting services LCCs do not have to worry about transferring baggage between aircraft and handling passengers that missed their flight (compensation) (Schlumberger and Weisskopf 2014).

LCCs directly connect the destinations that are usually served by traditional airlines through their hubs (indirectly). In this way, LCCs with their non-stop flights are focusing on the local origin and destination market and represent a by-pass for given destinations.

Most LCCs avoid congested airports and thus gain significant advantage over their rivals. LCCs primarily fly to secondary airports because these airports can provide certain benefits such as: less congestion, lower landing costs and short time for aircraft turnaround. All these benefits allow an airline to achieve high utilisation rates (aircraft and labour utilisation) (Schlumberger and Weisskopf 2014). With increased airport utilisation and point-to-point route structure LCCs manage to achieve more flights with less facilities and employees, which further reduces costs. LCCs combine more daily flights per aircraft and maximum number of seats in the passenger cabin that results in more produced ASK per aircraft. This ASK productivity contributes to lower unit operational costs of LCCs (Schlumberger and Weisskopf 2014).

By avoiding congested hubs, LCCs have managed to achieve great accuracy in the execution of their flight schedules, which has made them additionally recognizable on the market. They record the highest levels of punctuality and the lowest number of cancellations, denied boarding, lost bags, and customer complaints (ELFAA 2004). This made them very attractive not just for leisure, but for business passengers, too.

Secondary airports. The development of secondary airports is closely linked with the LCC business model. Most of Europe's secondary airports were built for military purposes, and later they were used as regional airports. However, the frequency of regional flights was very low, so for many years these airports were underutilised (e.g., Frankfurt-Hahn and Bergamo-Orio al Serio) (ELFAA 2004). This situation has led both sides, LCCs and secondary airports, to start to cooperate and gain mutual benefits.

Secondary airports allow LCCs to operate smoothly and achieve high efficiency. The availability of airport capacity throughout the day allows airlines to design optimal flight schedules in terms of maximum fleet and labour utilization. Due to the absence of congestion, they are enabled to perform their flight schedules on time and have no additional costs due to delays, and it also benefits in maintaining a large daily aircraft block-hours (ELFAA 2004).

Secondary airports benefit from LCC service in terms of increased passenger volumes which further leads to increase in both aeronautical and non-aeronautical[3] revenues and, ultimately, higher profitability. In order to maintain a certain level of traffic, secondary airports offer airlines lower fares for their aeronautical services, helping LCCs to maintain low operating costs. Due to the fact that these airports operated below their

3 Non-aeronautical revenues can be sourced from different commercial areas such as car rentals, shops, banks, post offices, restaurants, bars, etc. The airport gains rental and concession fees from these businesses, which tend to increase proportionally with the increase in passenger numbers. Moreover, an airport can gain charges for car parking or any other transport organization (e.g., shuttle-bus) to and from the main city centers they serve.

capabilities until the LCCs started to use their services, investment in their equipment was generally absent. This leaves the possibility for LCCs to design all additional equipment according to their specific needs. For example, airport terminals usually do not have enough counters to register passengers, and this allows them to design them as simply as possible so that these processes can be performed in the shortest possible time. Some airports designed the entire terminal for their biggest customers among LCCs where they also charged lower fees for passengers (e.g., Marseille and Geneva airports) (ELFAA 2004).

Aircraft utilisation. Another LCC business model attribute is high level of aircraft utilization. This goal is easiest to achieve in combination with the use of secondary airports and the operating of point-to-point route structure. The way in which LCCs increase the utilization of their aircraft is to have high frequencies of flights during the day, together with the reduction of aircraft turnaround time at airports (on average, their turnaround time is about 15–25 minutes) (Vasigh et al. 2018). By cancelling the reservation of seats, they manage to accelerate the boarding and disembarking passengers at an airport and, thus, remain on the ground much shorter than other airlines. With other services that are simpler because they are performed at smaller airports, achieving such a short aircraft turnaround time is very sustainable. Also, most of their flights are performed on secondary airports (avoiding the hub airports), which is why their total delay is very small, so that in the process of designing the flight schedule, they can plan a larger number of flights.

There are other strategies that some LCCs are using in order to achieve high aircraft utilization rate, such as flying longer routes (Jetblue and Easyjet) or increasing the number of flights per day (in average LCCs are utilizing their aircraft between 10 and 14 hours daily) (Dietlin 2004).

5.3.1 The impact of low cost carriers on the aviation market

Over the past four decades, LCCs have transformed the air transport market, shaking the traditional world of the network carriers, and creating a revolution in air travel service. They had an important impact on the passenger's choice of mode of transport, price reduction, the rapid development of regional and secondary airports, as well as the development of the regions themselves, increasing employment and popularizing tourism. There are a number of benefits in many areas that can be observed, but in this section only those that are most noticeable will be processed.

5.3.2 Impact on consumers

LCCs offered a wider choice of airlines and destinations to passengers at significantly lower prices on a large number of flights within one continent. Once the barriers of the free entry into the aviation market were removed, the number of LCCs has increased significantly. In the period before the deregulation of the aviation market, only traditional airlines had licenses to operate. In the EU this was an exclusive right assigned to national airlines. However, according to the European Commission, in the period from 1992 to 2000, the number of airlines increased from 77 to 139 (ELFAA 2004). Most of these airlines, which emerged after the liberalization of the market in the EU, apply a low-cost business model. Passengers in the EU, as well as in other open markets, are now able to choose between the various airlines as well as the services they offer.

Increased number of airlines on the market allow passengers to choose between different flight schedules, frequencies and airports where they want to fly. By servicing secondary airports LCCs offer passengers a greater choice. For example, one of the busiest routes in Europe, Dublin-London is currently served by Aer Lingus, CityJet, Ryanair and British Airways with flights to London Gatwick, London Heathrow, London Stansted, London City and London Luton. Dublin-London accounted for almost 4.5 million passengers on the route in 2015, Routes Online 2016. Ryanair is the largest airline by seats overall in Ireland, with a 40.2% share in 2019 (CAPA 2019).

The result of the increased number of LCCs on the market, the number of direct routes served has also increased, because their business model includes point-to-point flights. According to research by the European Commission in 2004, there were 30% more routes compared with the number of routes in 1993. Figure 5.2 shows the development of the number of direct routes served between 2001 and 2010 (ELFAA 2004). In particular, the number of routes served by Ryanair has increased exponentially (de Wit and Zuidberg 2012).

Tense competition in the air transport market has resulted in lower prices charged on a large number of flights. Maintaining low operating costs enabled these airlines to sell tickets at prices that were unimaginable in the period before deregulation. The great pressure that LCCs have caused on the market with low prices, has resulted in the reaction of traditional airlines to start reducing their prices on certain routes in order to remain competitive and keep their passengers (Dobruszkes 2006; Alderighi et al. 2012). In the Dublin–London market, in the period following the entrance of Ryanair in 1986 and up to 2000, traffic demand quadrupled while yields dropped to one-fourth (Franke 2004).

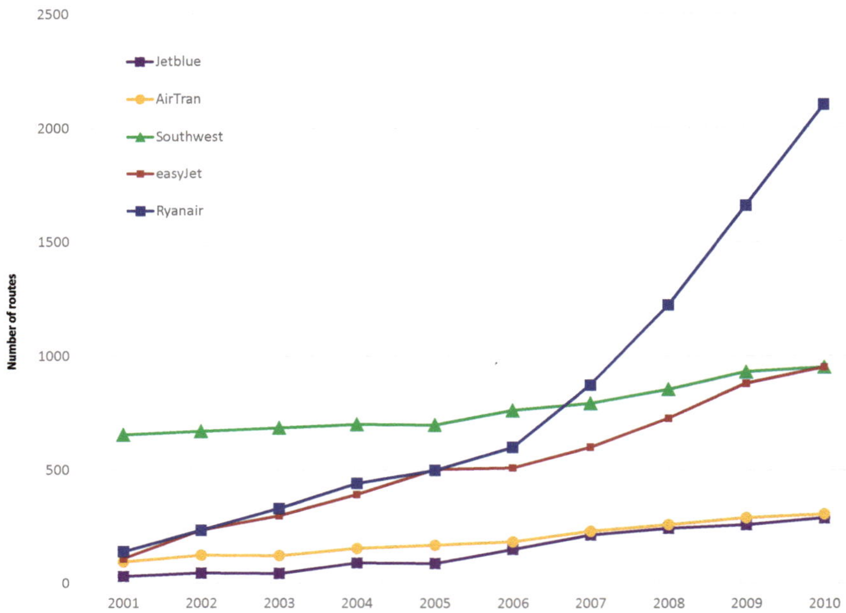

Figure 5.2. The development of the LCCs annual number of routes.

LCCs were first to offer the ability to book and buy tickets online. For many passengers this was an added value in terms of price transparency and simplification of buying tickets. Combined with extremely low prices, transparency in pricing has attracted a large number of new passengers. However, the success of this strategy is highly dependent on the degree of Internet penetration in a particular region (Schlumberger and Weisskopf 2014).

5.3.3 Impact on airports

Deregulation and liberalization of air traffic have brought many benefits to airports, too, because their development is closely linked to the development of LCCs. The biggest benefit is the increase in traffic, especially at regional and secondary airports, which have attracted LCCs (ELFAA 2004). Generally, before LCC entry at certain secondary airport, the number of operations at that airport was so low that most of the capacity remained unused. Due to such low capacity utilization, no airport slots are required to conduct traffic operations at that airport. Along with the low airport charges, this becomes very attractive for LCCs, since this allows to design flight schedules to make the best use in terms of fleet and labour (ELFAA 2004).

Given that the liberalization of air transport has forced airports to be more market oriented, many of them have entered into various negotiations with LCCs to create the most favourable conditions for traffic in exchange for attracting the passengers and opening new routes served from that airport. As airports have high fixed costs (construction and maintenance of infrastructure, equipment, staff salaries, security, etc.), no matter how much traffic they have, it is in their interest to maximize the number of passengers in order to distribute those costs (ELFAA 2004). A larger number of passengers at the airport leads to higher revenues that are generated from both primary (aeronautical) and secondary fees (parking, restaurants, shops, passenger transport to/from the airport, etc.). Increasing the number of passengers leads to enhancement in both revenues. The main difference between primary and secondary airports is that the later one do not see the aeronautical charges as their main sources of revenue but have oriented to commercial charges which include (ELFAA 2004):

- renting space in the terminal building for car rental agencies, restaurants, shops, etc.
- charging for parking the private vehicles,
- organizing transport by other modes of transport to populated areas,
- renting airport space for the purpose of advertising, etc.

The revenue from secondary activities may generate from 40% to 80% of all revenues at an airport (Oum et al. 2003). By increasing traffic operations and increasing revenues due to the emergence of LCCs, many airports have been able to upgrade existing infrastructure or expand their capacity to accommodate additional traffic and further increase their revenues.

5.3.4 Impact on regions

Many studies have shown that LCCs bring many benefits to both the regional airports they use and the regions themselves (ELFAA 2004, Williams and Balaz 2009, Whyte and Prideaux 2007). These studies indicate the impact of LCCs that they have on the development of tourism, employment and increased business activities in the region, due to the increase in the number of passengers and destinations at the local airport.

LCCs have increased the number of new tourist destinations to become accessible to a larger number of passengers at low fares. Also, tourism has become popular throughout the year, not just during the seasons, as these airlines have started offering scheduled flights throughout the year to attractive tourist destinations. Moreover, LCCs introduced a new type of tourism during the working days, by offering extremely low prices to popular tourist destinations outside the week-end peaks. This makes travel affordable for price-sensitive travellers who have extra time.

The emergence of LCCs has increased the employment both in the air transport industry and in the other industries in the regions that they began to serve. The employment generated by LCCs can be classified into three groups (ELFAA 2004):

- Airlines (pilots, cabin crew, engineers and administrative staff)—on average 166 of new jobs are created for every 1 million passengers in this sector.
- Airports, directly (administration, security, reception and dispatch workers, etc.) and indirectly (workers employed in secondary activities of airports, shops, restaurants, etc.)—in this sector, on average 950 of new jobs are created for every 1 million passengers.
- Employment in the region is increasing as a result of increased tourism activities in terms of opening new hotels, restaurants and other catering facilities. The economic development of the region and lower labour costs attracted many new businesses in other sectors, too. It is estimated that for every 1 million passengers, on average 1,100 of new jobs are created.

5.4 Charter airlines

Charter airlines are those airlines that operate charter flights, i.e., non-scheduled or ad hoc flights. Charter flights are flights organized on behalf of a group of passengers traveling together by aircraft to the same destination. These flights can be organized individually or by an agency (company) who rent the whole aircraft. They are usually sold as part of various travel arrangements and may include transport and transfer, accommodation and food (known as a package holiday), but can also be purchased individually (transport only, seat only).

Observing the historical development of the charter services (Trajković 1999), it can be noticed that the charter was individual in the initial phase of the development. Namely, a person rented a plane for his own, personal needs (for example transport of family, friends, or sometimes transport of employees organized by the employer). In the next phase of charter development, intermediaries who organize charter flights appeared. They can be airlines which intend to increase the volume of their scheduled traffic or travel agencies or tour operators which offer different arrangements within their tourism business. After this phase, the organizers of charter flights are faced with the problem of low passenger load factor in the leased aircraft; therefore, several operators participate in organizing one charter flight. Finally, in the last phase, charter flights are sold, as already mentioned, together with transfer, accommodation and meals within various tourist arrangements.

Charter airlines appeared in the 1960s. They offered flights in attractive tourist destinations, where they competed with traditional airlines. The main characteristics of charter airlines are that they operate their flights with a passenger load factor near 100%, while the level of service may be lower than the level of service in traditional airlines. Charter flights usually offer only one passenger class in the passenger cabin. Furthermore, the seat pitch is smaller in order to provide the possibility to offer a larger number of seats. Flight schedules are not coordinated with other schedules, i.e., they are performed point-to-point flights, and usually in off-peak periods when airports are not congested (night flights). Airport lounges are not available to passengers, and these airlines are not members of loyalty programs. When it comes to the frequency of charter flights, it can be said that it is small—charter airlines offer one or two flights during the week. As it is mentioned before, the characteristic of the charter is that the airline sells the entire flight, and the risk of charging is taken over by the flight lessee, i.e., travel organizer. All of the above cause the unit costs of charter are lower compared to scheduled traffic provided by traditional and regional airlines. Moreover, it indicates a lot of similarities with the low-cost business model.

Competition between charter airlines and LCCs. Airline market is a dynamic system, subject to various influences; therefore, in order to survive and to retain or even strengthen their market position, airlines have to adapt to the new conditions. With the advent of low-cost airlines, charter airlines have begun to operate on long-haul routes due to increased competition on short-haul and medium routes (Tatalović et al. 2012).

As already mentioned, charter airlines found out that their key to success is offering services on tourist destinations. In Europe, they were competitive with traditional airlines in these destinations with significantly lower transport costs. The entry of low-cost airlines into this market influenced the emergence of new competition between two business models—charter and low-cost. Due to very high competition in the liberalized market, charter airlines are beginning to focus on longer routes outside Europe where there is no competition from LCCs.

Comparing the characteristics of charter airlines and LCCs, certain differences as well as similarities can be observed (Kurth 2010). In the case of LCCs aircraft turnaround time is lower almost twice (25 minutes), flight duration is not higher than an hour and a half, additional services are charged, the fleet is homogenous, they sell the tickets online, they have developed marketing strategy, and operate point-to-point flights. In the case of charter airlines, aircraft turnaround time is higher (45 minutes), they offer flights on short and long-haul destinations, additional services are free of charge, the fleet is heterogeneous, they sell tickets through travel agencies, they have limited marketing strategy, and operates both point-to-point flights and hub and spoke. The common characteristic for charter airlines and LCC is that both of them offer flights at primary and secondary airports.

Summarizing all, it could be said that the main differences between these charters and LCCs come from different costs (Williams 2001), mainly from:

- Aircraft size,
- Distance length,
- Passenger load factor,
- Aircraft utilization,
- Labour productivity,
- Distribution costs,
- Passenger service costs,
- Landing fees,
- Administration and finance expenses.

Each of these elements will be described in more details.

Aircraft size. Regarding the aircraft size in the fleet, it can be noticed that charter airlines have aircraft with different capacities in their fleet. On shorter and medium routes, they mainly operate smaller capacity aircraft, while for long-haul routes, they employ aircraft with a capacity of over 300 seats (the average aircraft capacity at Tui Airways, one of the largest charter airlines in January 2019 was 233 seats). On the other hand, low-cost airlines generally have aircraft with a capacity of up to 200 seats (e.g., Ryanair has aircraft in its fleet (B737-800) with a capacity of 189 seats, the average capacity of Wizz Air's fleet, which includes two types of aircraft, both A320-200 and A321-200) is 196 seats, while the average capacity of easyJet's fleet (also two types of aircraft—A319-100 and A320-200) is 172 seats). As the size of the aircraft affects the costs of the airline, it is evident that the charter airlines have lower direct operating costs per passenger kilometre due to the aircraft with higher capacity in their fleet.

Distance length. Considering the distance length, it can be seen that, on average, charter airlines operate on longer routes compared to low-cost ones. The average distance length of charter airlines is about 2000 km, unlike low-cost ones whose average distance length is 1000 km. The average distance length directly affects:

- fuel costs per block hour—which are lower on longer routes; namely, cruising phase is the largest part of long-haul flights and in that phase fuel consumption is lower compared to other phases of flight;

on the other hand, in short-haul and medium-haul flights the share of cruising phase in total flight is smaller and accordingly fuel costs increase;

- maintenance costs—which decrease with increasing distance length, as aircraft maintenance and replacement of individual parts depends on the number of take-offs and landings.

Passenger load factor. Passenger load factor is one of the indicators of the use of the offered capacity. It can be noticed that charter airlines and LCCs have very similar load factors, although at the beginning of the 21st century the load factor of charter airlines was higher compared to LCCs.

Aircraft utilization. Aircraft utilization in the case of charter airlines are larger because they operate on longer routes, but also because they operate at night, while LCCs are limited to a certain period during the day (scheduled traffic); LCCs fly on short and medium routes, therefore, the utilization is lower due to the multiple turnaround time between many short flights.

Labour productivity. Cabin crew at charter airlines have a larger number of flying hours compared to LCC's cabin crew but they are paid less; many of them are not permanently employed, but only during the season, which further reduces the labour costs.

Distribution costs. Charter airlines do not have distribution costs, due to the fact that tickets are sold and promotions are performed by tour operators with whom they cooperate.

Passenger service costs. Passenger service costs are mainly related to in-flight catering and entertainment, and they are minimised. All these services represent an advantage for LCCs due to they are not provided free of charge (additional source of revenue). Bearing in mind that passengers need to pay for refreshment it is expected that refreshment will be less consumed (passengers generate less waste), which further influences on reducing the time needed for cleaning the aircraft, and finally to the decrease of aircraft turnaround time.

Landing fees. Charter airlines fly on longer routes, which indicates lower frequency and lower number of take-offs/landings, as well as accordingly lower landing fee costs. Additionally, landing fees are often lower due to the avoidance of peak hours at airports, and usage of secondary, less congested airports.

Administration and finance expenses. Charter airlines have lower administrative costs because the administrative tasks of charter airlines are most often performed by their tour operator partners.

Comparing charter airlines and LCCs it can be observed that LCCs offer wider variety of departure times, length of stay at the destination, as well as possibility for self-organization of the trip. In the case of charter airlines, it is harder to choose desired departure times, length of stay at the destination due to the fact that charter flights are usually a part of the package holidays.

The competition between charter airlines and LCCs is perceived on shorter routes (flight duration less than 2.5 hours), while on longer routes (flights lasting more than 3.5 hours) charter airlines have almost no competition. For example, from some destinations originating from Europe to European tourist destinations such as Salzburg (Austria), Grenoble (France), Greek's islands, Italian coast, Spanish islands, etc., the competition is extremely strong. On the destinations to Turkish or Egyptian holiday resorts the competition is weak, while charter airlines are without completion on long-haul routes (from Europe to Cuba, Jamaica, Mexico, etc.).

5.5 Cargo airlines

Cargo airline is an airline that transports cargo by air. There are four basic types of airlines that provide these services:

- Integrated express operators—operate a fleet of scheduled aircraft, trucks, and couriers offering door-to-door delivery service;
- All-cargo airlines—dedicated to air cargo transport;
- Scheduled passenger airlines—transport freight as a way of filling up spare capacity on their passenger aircraft;
- Cargo charter airlines—unscheduled air charter operators who transport freight from airport-to-airport.

The air cargo transport service provides many advantages and disadvantages compared with other modes of transport. The benefits of air cargo are its reliability, high standards of safety, security and speed especially over long distances. The other side of coin is that air cargo transport is the costliest cargo forwarding solution, typically priced 4–5 times that of road transport and 12–16 times that of sea transport (World Bank Group 2009). Its higher rates make it unsuitable for goods with lower cost of production. Moreover, the operating aircraft capacity is very limited that in turns provides lower overall transport capacity compared to maritime or rail transport.

5.5.1 The demand in air cargo market

Cargo airlines are faced with more options than in passenger transport in terms of definition of the major markets for air cargo services. Since the emergence of cargo airlines, the market they have dominated has been the market for the transport of high-value goods, emergency cargo, perishable goods, and goods that required special security conditions (Doganis 2010). Each of these segments requires certain conditions which can be provided by air transport and are significantly better in comparison to other modes of transport, such as higher delivery speed, increased security in terms of reduced damage and loss, etc.

The transport of emergency cargo arises when the goods need to be transported in a short period of time by the fastest mode of transport and when the cost of transport is relatively unimportant. The examples of such situations may be when companies need to correct an operational problem (an oil company must suspend work on a well due to equipment failure), or when there is a breakdown on an overseas ship and spare parts need to be delivered urgently. Also, the medicines such as vaccines are often carried as an emergency good due to their specific requirements. Another type of emergency is market-related urgency. For example, when suppliers are in danger of exceeding the delivery deadline or suddenly receive a large order from their essential customer. In this case, maintaining customer loyalty comes first.

This market segment is very important for airlines if they want to become competitive. In order to achieve that, the most important thing is to offer a high frequency of flights due to the unpredictability of emergency situations (Doganis 2010). This type of order must be processed immediately and must offer precisely defined transit times in order to meet the tightest of deadlines. Due to the irregular and unpredictable nature of this type of demand the available storage space is very important in order to accommodate the goods of larger volume and size. Moreover, an airline has to have the appropriate equipment for handling different types of goods in terms of volume and size. Additional services such as access to important documents like air waybills, proof of delivery and tracking system for shipment at all times must be met, so that the air transport service becomes competitive with other transport modes.

The goods characterized with a high-value in relation to its size and weight are another important segment in the air cargo industry. This type of good is usually gold, diamonds, museum artefacts, etc. High-value goods have always required extra security, i.e., a range of special procedures and security to protect the cargo from unpredictable incidents throughout every step of the transport process (from the warehouse, to transit and even at highly secured airport areas). Thus, the transport of valuable goods presents many unique challenges to airlines and logistics providers compared with other types of air cargo. Cargo airlines dedicated to transport the valuable goods should be focused on designing optimal routes in order to identify and avoid risk areas. Also, it is important to monitor and measure security procedures to ensure the safest route with minimal exposure to risk through the supply chain. Along with high security standards, the speed is important in valuable goods transport. This is the reason why the air transport is extremely attractive for this market (Hsu et al. 2013).

The basic requirement for airlines in transporting the perishable goods is to transport the goods within a certain period of time, while it is still saleable at the market (Shaw 2020). There is a difference between physical perishability (e.g., fresh flowers, fresh fruits and vegetables, seafood, etc.) and economic perishability (e.g., books, newspapers, seasonal clothes, etc.) of goods. Economically perishable goods are those types of goods whose time period in which they have a value is very short. This type of goods can be sold in large quantities and at high prices if it arrives on the market on time. The transport of perishable goods differs from the transport of emergency cargo in terms of predictability of the occurrence of this type of transport requests. It is possible to forecast when the need for transport of these goods will occur since the production of most perishable goods is seasonal. The main challenge for airlines that transport the perishable goods is the determination of transport price, because the air transport service prices tend to be high relative to the value of this good (Shaw 2020). There are certain limits to which customers are willing to pay these high prices. Namely, the economy of exporting perishable goods is based on the basic profit realized in more distant markets, which covers additional costs of transport by air and leaves a certain profit. If the price of transporting the goods by air exceeds the potential earnings, the request for the service will be cancelled.

Express shipments cover the transport of small, urgent packages and represents a very important segment for the development of air freight transport. The courier, express and parcel (CEP) market is very time sensitive and cannot wait for long deliveries, sometimes not a single day. Airlines that operate passenger traffic are not very useful in this case, because the passenger traffic is usually done during the day. Express cargo airlines generally offer a full service which includes picking up goods from the sender at the address, delivery of goods to the end user, also at the specified address, transport service, the ability to track shipments, invoicing and procurement/processing of all necessary papers. Although express delivery makes a small part of the total air cargo volume, the revenue generated through this type of service makes proportionally a larger part of the total revenue of the cargo airline industry (Mazareanu 2020).

However, the greatest part of the shipments moving internationally are non-perishable goods and cargo airlines must be competitive in this segment as well, otherwise they will not use all their potential. Namely, the other modes of transport are much more favourable for the client in this segment of cargo industry, because there is no need to transport the goods in the fastest possible way. Customers who still choose to transport the goods by air are aware of the fact that it is more expensive than any other mode, but also want to take advantage of greater market flexibility (e.g., to be the first to offer goods on the market) (Shaw 2020). Also, air transport offers other benefits for transporting non-perishable goods such as lower documentation, packaging, insurance and warehousing costs, and in many cases these savings can offset the higher transport costs (Doganis 2010).

Freight forwarders have a big role in creating a market for the air transport of goods. This is mostly related to the pricing of air cargo. Cargo airlines determine the price for transport of goods based on the principle of "discount price per kilogram of goods, the heavier the goods", and this gives freight forwarders many opportunities. They collect smaller orders and form one large one which they then hand over to cargo airlines. Beside the fact that airlines try to stimulate demand with the special prices, other justification for that they see in the reduced handling cost and cost for preparing the documentation, where these costs are lower for larger shipments (Doganis 2010). Therefore, freight forwarders can group a larger number of smaller shipments into one larger and thus hand it over to the airline, which will charge them for transport at a lower special price. The freight forwarder will transfer a part of that lower transport price to the actual senders, who will therefore pay a lower price compared to the one they would pay if each individual had given his/her shipment directly to the airline. The part of the price difference freight forwarder keeps for himself in order to secure his profit.

5.5.2 Cargo airline network and flight schedule

A large number of airlines transport goods in the belly-holds of their passenger aircraft. In that case, they do not have separate networks and flight schedules for the transport of goods and passengers. The main disadvantages of this practice are reflected in the limitation of the offered capacity for accommodation of goods as well as in the flight schedule that is designed according to the requirements of passengers (Holloway 2008). Airlines that are dedicated to transport goods or those that perform combined transport, but have separate fleets for passengers and goods, have more flexibility in designing the route network and flight schedules.

The network design of cargo airlines is predominantly linear, hub-and-spoke and point-to-point. The hub-and-spoke network structure enables the efficient development of a global network coverage by maximizing the number of destinations under the restrictions of the airline's capacity. The point-to-point network connects the airports with direct flights. Direct flights offer the reduced total travel time and thus are less sensitive to the occurrence of flight delays. The flights are evenly distributed among the airports, no airport has a large number of concentrated flights, which reduces the possibility of disruption and the spread of disruption to other flights in the network.

The flight schedule of cargo airlines differs from the flight schedule in passenger traffic. Freight forwarders prefer transport services to be performed during the night, especially when it comes to express shipments, so departure times are generally scheduled to their requirements. This gives freight forwarders the opportunity to collect the shipments during the day, and then at the end of the day to give those shipments to the airlines. Transport of shipments is done during the night, so that the goods arrive at the end user early in the morning (depending on which time zones are crossed and how urgent the delivery is).

5.5.3 Airline cargo fleet

Cargo airlines tend to use new aircraft, especially designed to transport cargo, known as a freighter aircraft. The main reasons behind this strategy is to achieve maximum payload-range capabilities for a given aircraft size (freighter has 50% higher payload compared with the same aircraft in a passenger configuration) and to overcome the handling problem with large and oversized shipments (freighters usually have larger doors) (Doganis 2010). Some of the examples of the freighter aircraft are Boeing 747 cargo, Boeing 767 cargo, Antonov 124–100, etc.

Also, there are many cargo airlines that use old aircraft which have been withdrawn from passenger traffic and adapted for the purpose of transporting goods. This practice is adopted in order to reduce costs. The most common types of aircraft that are adapted for transporting goods are Boeing 707, Boeing 727, Douglas DC-8, Ilyushin Il-76, etc. In order to meet all the conditions necessary for transporting the cargo, these aircraft are equipped with special equipment for handling and fixing pallets and containers, windows are closed with opaque materials, their floors are strengthened, etc. However, as the maintenance costs of these aircraft tend to increase, this practice became very costly. Also, there is the noise factor they produce, which is becoming increasingly unacceptable at most airports in the world.

The following characteristics are important for freighter aircraft (Holloway 2008):

- Design density—represents the density that combines the full use of volumetric capacity with the maximum possible payload. This feature has been introduced in cargo transport due to the aircraft limitation in terms of volumetric capacity and as the payload.
- Door dimensions—it is important because it determines how easily containers, pallets and other oversized goods will be loaded into the aircraft.
- Fuselage cross-section—determines the size of the containers which can be placed accommodated, as well as the volume of space above the standard containers, which can be used to accommodate additional cargo.
- Weights of aircraft parts—such as doors, strengthened floors and other equipment inside the aircraft for cargo handling, because it increases the overall aircraft weight and reduces the payload.

5.5.4 The price determination in transporting goods by air

In the period before air transport deregulation, international cargo rates were regulated at the international level within the IATA framework and afterwards approved by national governments. After deregulation, the cargo rates on many international routes has become very variable and subject to the influence of current market conditions. It is noticeable that there is a pronounced competition among airlines based on the cargo rates, especially in the part of the market where forwarders operate. The standard tariff structure of cargo airlines consists of the general retail rates and contract rates. The cargo rates are usually expressed as a rate per kg (lb) and the principle that the rate per kg decreases as the size of the shipment increases is applied.

On many routes additional tariffs may be available defined in terms of quantity (quantity rates), specific characteristics of goods (specific commodity rates) or particular unit load devices used (ULD rates). These additional tariffs are introduced to stimulate demand by encouraging consolidation into large shipment, attracting specific types of goods and using unit load devices that will decrease airline handling costs. Moreover, general cargo rates may include cargo tariffs which involve surcharge on the basic rate in respect of certain categories of goods. For example, surcharges are applied for carrying live animals, human remains or other goods that request special handling services.

Contract rates are lower than general retail rates and airlines usually apply them to individual freight forwarders with significant tonnage of goods for transport. Most cargo airlines rely heavily on freight forwarders that guarantee them sufficient amounts of goods and capacity utilization over a given period.

The increase in average cargo rates in recent decades is mainly related to the improvement of the cargo service from airport to airport. Many cargo airlines invested in distribution networks, in electronic data interchange and in other facilities which are important from the customer's perspective (Doganis 2010). Moreover, integrated operators (e.g., FedEx, UPS and DHL) offer a seamless door-to-door services that include collection, transport by air or truck, distribution and just-in-time services. Their "overnight" service, second-day and third-day "time-definite" service, is certainly changing the dynamics of the air cargo industry significantly.

5.5.5 Balancing the payload and volume in determining the cargo rate

When determining the cargo rate in air cargo transport, it is important to optimize the payload-volume ratio in order to maximize the aircraft capacity utilization and revenue. As mentioned earlier, each aircraft has a fixed volumetric capacity that can be used in transporting the goods and a payload capacity that varies depending on weight of fuel and the number of passengers (in the case of combined transport) for a given flight (Holloway 2008).

This is why airlines have a problem in determining what service charges to apply, based on weight or based on volume of the good. It is clear that it makes no sense to fill the available aircraft space with light goods, and then charge the transport service per unit of weight, or in another extreme case, reach the maximum payload of aircraft with high density goods but small volumes and then charge the transport service per unit of volume. In order to overcome this problem and to achieve a certain level of balance between payload and volume airlines apply a charge based on assumed weight per unit of cubic capacity (so-called dimensional weight) (Holloway 2008).

References

Alderighi, M., Cento, A., Nijkamp, P. and Rietfeld, P. 2012. Competition in the European Aviation market: the entry of low-cost airlines. Journal of Transport Geography 24: 223–33.

Berry, S., Carnall, M. and Spiller, P.T. 1996. Airline Hubs: Costs, Markups and the Implications of Customer Heterogeneity (Cambridge: National Bureau of Economic Research).

CAPA. 2019. Ireland aviation: good geography, good policies. https://centreforaviation.com/analysis/reports/ireland-aviation-good-geography-good-policies-462870. Accessed on April, 2021.

De Wit, J.G. and Zuidberg, J. 2012. The growth limits of the low cost carrier model. Journal of Air Transport Management 21: 17–23.

Dietlin, P. 2004. The potential for low-costs in Asia. Massachusetts Institute of Technology.

Dobruszkes, F. 2006. An analysis of European low-cost airlines and their networks. Journal of Transport Geography 14(4): 249–64.
Doganis, R. 2010. Flying off Course: Airline Economics and Marketing. Routledge, 4th ed.
ELFAA. 2004. Liberalization of European air transport: The benefits of low fares airlines to consumers, airports, regions and the environment. Brussels: European Low Fares Airlines Association.
ERAA. 2021. https://www.eraa.org/about/about-era/about-era/our-sector, Accessed: 17.02.2021.
Fageda, X. and Flores-Fillol, R. 2012. Air services on thin routes: Regional versus low-cost airlines. Regional Science and Urban Economics 42(4): 702–714.
Fischer, W.J. 2018. Transportation economic regulation in practice. TR News, 315: 3–8.
Franke, M. 2004. Competition between network carriers and low-cost carriers. Journal of Air Transport Management 10(1): 15–21.
Holloway, S. 2008. Straight and Level: Practical Airline Economics. Ashgate Publishing Limited, England, Ashgate Publishing Company, USA, 3rd ed.
Hsu, C.I., Chen, Y.H. and Chen, W.T. 2013. A study on airlines differentiated cargo service strategies. Transport Policy 25(1): 101–110.
Krstić, B. and Stepanović, L.J. 2004. Avio-saobraćaj u turbulencijama: neki elementi za formiranje vazduhoplovne politike, Centar za liberalno-demokratske studije, Beograd.
Kurth, W. 2010. Low-cost Carriers -Threats and Opportunities, Hellenic Chamber of Hotels, Athens.
Mazareanu, E. 2020. European courier, express and parcels market - statistics & facts. Statista. Published on August 31, 2020.
Oum, T., Zhang, A. and Zhang, Y. 2003. Concession profit and its efficiency implication on alternative forms of economic regulation of airports. Seventh ATRS World Conference, Toulouse.
RAA. 2020. https://www.raa.org/content-hub/raa-annual-reports/raa-2020-annual-report/.
Routes Online. 2016. Dublin - London now the busiest city pair in Europe. https://www.routesonline.com/news/29/breaking-news/253012/dublin-london-now-the-busiest-city-pair-in-europe/. Accessed on April, 2021.
Schlumberger, C.E. and Weisskopf, N. 2014. Ready for Takeoff? The Potential for Low-Cost Carriers in Developing Countries. Directions in Development--Infrastructure. World Bank Group, Washington, DC.
Shaw, S. 2020. Airline marketing and management. Routledge; 7th edition.
Tatalović, M., Mišetić, I. and Bajić, J. 2012. Airline management, MATE d.o.o. Zagreb (in Croatian language).
The World Bank Group. 2009. Air Freight: A Market Study with Implications for Landlocked Countries. World Bank's International Trade Department, The World Bank Washington, D.C.
Trajković, M. 1999. International aviation law, University of Belgrade – Faculty of Law and Jugoslovenski aerotransport (in Serbian language).
Vasigh, B., Fleming, K. and Tacker, T. 2018. Introduction to Air Transport Economics: From Theory to Applications, Routledge, 3rd ed.
Wensveen, J. 2015. Air Transportation: A Management Perspective. Routledge, 8th ed.
Whyte, R. and Prideaux, B. 2007. Impacts of low cost carriers on regional tourism. Paper presented at the CAUTHE 2007: Tourism e Past achievements, future challenges, Sydney, Australia.
Williams, G. 2001. Will Europe's charter carriers be replaced by "no-frills" scheduled airlines? Journal of Air Transport Management 7: 277–286.
Williams, A.M. and Balaz, V. 2009. Low-cost carriers, economics of flows and regional externalities. Regional Studies 43(5): 677–691.

Chapter 6

Airline Planning and Operations

6.1 Airline demand

Air transport demand potential can be defined as the number of potential passengers, those who desire the service but also have the ability to pay for the air service on a particular market or markets, with assumed conditions of the service offer. This demand could be satisfied or unsatisfied (Holloway 2008). Satisfied and unsatisfied demand together represent the potential market. Satisfied demand is passengers who have chosen air service on the given market, while unsatisfied demand is passengers who have chosen either air service from another airport, or some other transport mode, or have decided to give up on the intended trip. Therefore, the airline has the air transport market share in the actual market, while the real air transport market (the actual market) is part of the potential unused market (Figure 6.1). Increasing the origin-destination distance of a particular journey reduces the competition of other modes of transport. This leads to a larger share of the air transport market in the potential transport market.

Passengers of various segments are consumers of services produced by airlines, and they represent the demand for services provided by airlines. Demand and supply affect each other. It is clear that the response to an increased demand on the air transport market will be an increase in supply (additional destinations, increase of flight frequencies, etc.). On other hand, additional supply can also induce demand. For example, introduction of low-cost service can have an effect on attracting passengers from the area who currently use other modes or who did not travel at all (attracted by very low price). If the new service is 'better' than the services provided by other airlines at the given airport or the service at

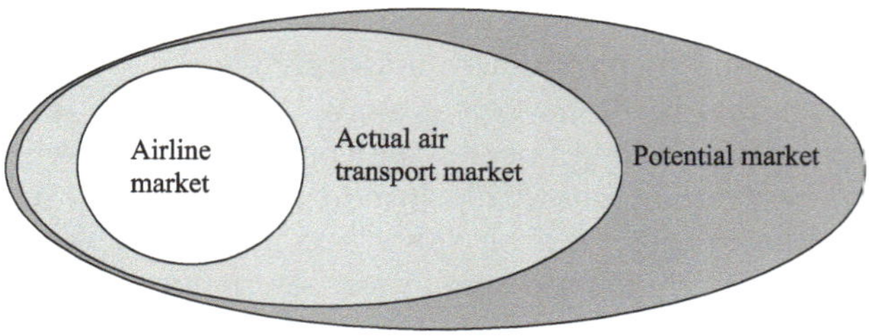

Figure 6.1. Potential market, actual air transport market and airline market.

neighbouring airports (destinations, frequencies, or even fares), the airline will attract travellers, thus further increasing the demand.

The air transport market is characterized as a highly competitive market where airlines are under great pressure to deliver services of desirable quality. The increasing competition is reflected through prices, airline schedules, and networks, different levels of services, and a growing number of airlines offering their services on a specific route and on the whole market. The airline sector is a part of the transport market, where margins are very small. Thus, airlines try to find strategies that allow them to become more efficient, mostly by applying economies of scale either through the size of their own company or by cooperating with other airlines in alliances (Bieger and Wittmer 2011). Through adequate air transport supply, airlines tend to attract as many passengers as possible.

Generally, air travel demand is driven by socio-economic activities, expressed by indicators related to GDP, trade, population, tourism, etc. Taking into account the definition of the derived demand, which is the demand resulting from the demand for different or related goods or services, it follows that air travel demand is derived. The air transport demand is based on the desire of passengers to be somewhere else for different reasons (leisure or business), not just to fly by plane.

Air transport demand is affected by service-related (a), socio-economic (b), and other (c, d) factors, such as:

a. air ticket price, service quality (reliability), the timing of flights, daily flight frequency, non-stop flights or flights with stops, comfort related to an aircraft seat, on-board service and catering, entertainment options, airport accessibility, and related travel expenses;
b. passenger demographics (population, age distribution, income levels), preferences of passengers; the number of days of holiday available for travel or leisure, and social attitudes towards travel;
c. the prices of competitors, availability of other modes of transport;
d. random factors (disease, terrorism, or natural disaster).

Based on the above mentioned demand drivers, it is necessary to consider the demand at both macro and micro-levels. There are factors affecting the demand in all markets. Among these general factors, the air ticket price and distribution of personal income on the markets served are the most important (Doganis 2010). In addition, there are several other factors that influence demand on some specific routes, such as tourist attractiveness of particular destinations, exchange rate fluctuations or even political factors (for example, visa restrictions make mobility and travel difficult for travellers from certain countries), historical and cultural relations between two countries, earlier population movements, migrations, and present labour mobility. Emigrants who visit their homeland, as well as their friends and relatives, who visit them, constitute a significant portion of air travel demand.

The air transport demand is characterized by several 'negative' features such as fluctuations, cyclicality, seasonality and peaking, and directional flow. A fluctuation or demand oscillation is caused by a number of different factors that can influence demand. Cyclicality is related to a long-term trend of peaks and troughs in air travel demand, which is the consequence of periodical changes in the economic activity level (recession and growth). Seasonality refers to short-term variations in demand, where demand increases during the summer months and then declines during the winter months. Thus, charter airlines operate intensively during seasonal peaks in the holiday season. In addition, a short-haul market is characterized by weekly and daily peaking (Monday, Friday, weekend as well as morning and afternoon peaks during the day). Directional flow is an increased demand in one direction for a given period of time, while in the opposite direction aircraft fly relatively empty (sports events, world cups, Olympics games, etc.).

Generally, demand can vary according to the reason for travel. Drivers of business trips are national economic development, international trade, growing industrial globalization, and business expansion strategies, while non-business travel is a function of increasing social interaction, level of education (better education leads to a desire for travel experience, length of holidays, the availability of attractive and acceptable tourist facilities, the ability to travel in financial and political terms, etc.) (Shearman 1992).

6.2 Airline supply

Characteristics of the transport supply depend on: technology, operational strategy, institutional requirements and restrictions, and users' behaviour (Kanafani 1983). Technical characteristics of the transport system affect its performance. Operating costs depend on the type of technology used. Also, speed and capacity directly depend on the technology. The way in which the technology is used to provide the transport service depends on the company's strategy and the way it controls its business. Finally, the behaviour of transport system users affects certain aspects of the transport supply.

Airlines must have a strategy based on demand analysis and forecasts, to attract and fulfil expectations of different segments of potential users—air passengers. Thus, airlines tailor different types of transport services based on the market research and the requirements of two opposite segments of users (business and non-business travellers).

Non-business travellers are offered the service at a lower price; namely, air ticket reservations can be made months or weeks in advance (before the trip) but with very limited flexibility (fare restrictions). Unlike non-business travellers, business travellers are offered the option of booking at the last minute, as well as great flexibility in terms of changing the date or time of travel of an already purchased ticket. However, this type of transport services is charged significantly more in comparison to the price paid by non-business passengers (within each tariff, the prescribed conditions apply, high tariffs give great flexibility in travel plan changes) (Tretheway and Oum 1992).

Moreover, airlines must monitor the operations of competing airlines on the observed market. Finally, airlines want to maximize their revenue and profits, in both the short and long term. Therefore, when planning a product that will be offered to the air transport market airlines take into account market segmentation, competing airlines and operating costs.

Product planning is important for two reasons. First, it has a key role in the process of matching potential air travel demand with the actual supply offered by airlines. Second, product planning has a direct impact on operating costs (Doganis 2010).

According to Doganis (2010), the airline's potential users will be affected by five key product features when making travel decisions and choosing between airlines:

- price, including fare levels and conditions,
- airline schedule (points served and routings, frequency, timings, connections and punctuality),
- comfort (type of aircraft, interior configuration, individual space, on-board service, ground/terminal service, airline lounges and in-flight entertainment),
- convenience (distribution/reservations system, capacity management policy, seat availability, ability to change reservations),
- image (reputation for safety, branding, frequent flyer loyalty schemes, promotion and advertising, market positioning).

The airline's supply refers to the airline's ability and willingness to provide a certain number of seats at certain prices in a given period of time and on a given market. Supply is the capacity of an airline to transport passengers and it depends on the network—destination served and availability of aircraft (Vasigh et al. 2018). Each airline tends to maximize profits through supply using price and income elasticity. Therefore, matching demand and supply is essential for success in the aviation industry.

According to Vasigh et al. (2018), factors that influence supply are:

- air ticket price,
- price of resource inputs (fuel prices, labour costs, aircraft and maintenance costs, airport charges, etc.),
- navigation charges,
- technology,
- availability of other modes of transport,
- government regulation, and
- stochastic factors (weather conditions, strikes).

Three main characteristics of the airline's supply are seasonality, rigidity and perishability of the airline's product. As already mentioned, demand is characterized by annual seasonality, which results from tourist movements in certain periods. Therefore, airlines must respond to this variability in passenger demand by increasing the frequency of flights on certain routes, in terms of introducing new flights. To match supply and demand, the airline must either withdraw certain capacity from the existing routes or, if it has available capacity, allocate new additional flights. In both cases, there is an increase in costs, but also in income.

Another characteristic of the air transport supply is rigidity. It is not easy for an airline to react to unexpected changes in demand in a short period of time. Namely, all traditional airlines and airlines that operate on a scheduled traffic design publish seasonal flight schedules twice a year, which define the lines that airlines will fly, on which days, at what time during the day there are take-offs/landings, type of aircraft, etc. The possibility of booking airline tickets in advance, resource planning (aircraft owned and leased aircraft), as well as contracts with employees do not provide the possibility of supply flexibility in terms of responding with another offer in a short time.

The unique characteristics of airline supply are the perishability of the product and the absence of inventories (air services cannot be stored). This means empty seats are worthless once the plane takes off. In other words, an airline can only sell airplane seats prior to the departure.

6.3 Market segmentation

Air transport market segmentation is the process of dividing passengers into segments based on trip purpose and their requirements related to the airline's product, i.e., air transport service. Namely, a specific market segment means a group of air transport service users, who have enough in common, such as similar behaviour, expectations and perceptions related to supply attributes, such as transport price, fare flexibility, flight schedule, flight frequency, and quality and level of transport services. The characteristics of different market segments and their respective sizes are very important inputs to determine the appropriate airline products. Without detailed knowledge of the market where airlines provide the service, it is not possible to match supply and demand in a proper way.

The question that arises is how to do market segmentation properly, taking into account that mistakes are often made with under segmentation or over segmentation. Under segmentation occurs when passengers are grouped into a small number of segments that are too large (numerous), and in which there is a high degree of diversity in requirements and expectations. On the other hand, over segmentation means a large number of segments that give results that cannot be used as inputs in policy development (Shaw 2004).

Depending on the objective of specific research, passengers should be divided into an appropriate number of segments. For example, a broad segmentation must be used for product planning purposes (economy, business and first class), while for marketing campaigns, passengers should be divided into much more segments (Shaw 2004).

Traditionally, the air passenger market segmentation is based on:

- the trip purpose,
- the trip length (flight/s duration) and
- the passenger origin, taking into account the heterogeneity related to the country of origin and their culture).

6.3.1 Trip purpose

According to the trip purpose, passengers are basically divided into two groups: business and leisure travellers. But, it should be noted that there are trips that cannot be classified into one of these two groups, as, for example, pilgrims visiting Islamic shrines in Saudi Arabia or, individuals who go for medical treatment. These market segments should be considered separately. Despite the fact that there are exceptions, most trips can be classified into two groups: business and non-business trips.

Business and non-business travellers have different travel motives, dissimilar requirements, divergent behaviour and expectations. This is reflected in all aspects of travel planning, from booking and buying the ticket, changing travel times and/or routes, the need to stop the trip for one or two days (stop-over), the ability to pay a certain fare, to the required comfort and level of services on board and at the airport.

Business travellers are customers who are not sensitive to travel costs (usually the company covers the costs), but they are very demanding when it comes to travel time (direct flights or flights with one or more stops), take-off/landing time, return on the same day or return on Friday night (to spend the weekend at home). Business travellers can be divided into two groups, those who try to spend the minimum necessary time on a business trip and those who connect their business trip and a short vacation at a tourist destination. Furthermore, business trips can be organized at the last minute (which means that the purchase of the airline ticket is done one or two days before the trip) or they can be planned in advance. According to Doganis (2010), the business segment may be subdivided into routine business, emergency business, conference or trade fair travel, overseas employment, home leave, and so on (Figure 6.2).

Also, there are business travellers who are employed in corporations (large companies with a defined business travel policy) and independent business travellers (self-employed or who work in small companies). In the first case, travel expenses are covered by the company in which the business travellers are employed, and in the second case, travel expenses of individuals (who are aware of the level of costs in relation to the company's capabilities) are covered by their small company or by themselves. Although business travellers are not sensitive to travel costs, the segment of independent business travellers has an accurate sense of how much the price of an air ticket affects the company's budget. In any case, both groups of business travellers (corporate and independent) are interested in high flight frequency and punctuality (Shaw 2004).

The non-business segment includes leisure trips organized by a travel agency, self-organized leisure trips, visits to friends and relatives (VFR), emigration related travels, student travels, etc. (Figure 6.2). Leisure trips can be very different, depending on the length of stay at the tourist

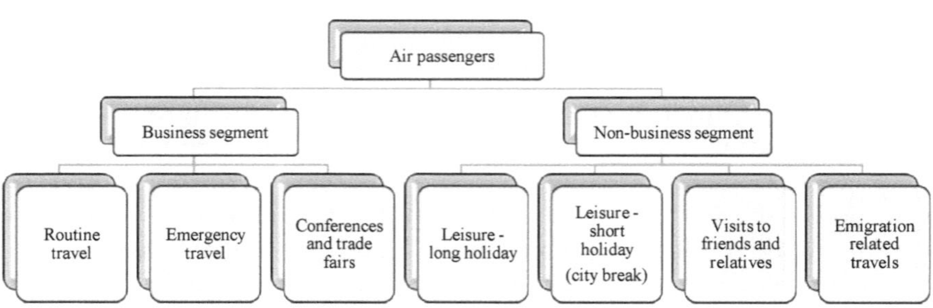

Figure 6.2. Air passenger segments and sub-segments.

destination. For example, these are one or two-week trips on the one hand, and very short, weekend trips (city breaks), on the other. Unlike business travellers, non-business travellers are very sensitive to travel costs. Thus they plan their trip in advance. These are often family trips, so in that case the costs of the trip itself and the stay at the desired destination are carefully planned.

As for passengers visiting friends and relatives, unlike typical tourists, these travellers do not pay for accommodation and other expenses related to the stay. Airlines pay special attention to this market segment because this is a group of passengers with permanent ethnic ties to emigrants. It should be noted that non-business passengers do not care about the daily flight frequency, travel time, and/or the number of stops.

6.3.2 Trip length (flight duration)

Requirements of passengers who are flying short-haul compared to those who are flying long-haul are very different. In the case of flights lasting from 45 minutes to one hour, the quality of airport service, including airport access and the experience related to the given airport, is very important for passengers. In the case of long-haul flights, the aspect of on-board service and time spent on the aircraft is very important to passengers. Thus, seating comfort and food served during the flight are crucial for passengers.

6.3.3 Passenger origin

Airlines pay attention to the origin of passengers, in terms of their culture and expectations, which may vary for passengers from other parts of the world. For example, there is a cliché of the business traveller in North America and Western Europe, who is middle-aged, well dressed and who carries a small amount of baggage. Business travel for other countries can have a completely different meaning, and the expectations of these travellers can be significantly different (Shaw 2004).

There are also significant differences between races, in terms of physical, behavioural and cultural attributes. On average, travellers from North America and Europe are taller and larger than people from the Far East. Therefore, the issue of comfort related to the size of the seat and the seat pitch and legroom is evaluated differently. Further, it is common to have a selection of hot meals on-board, especially in the case of long-haul flights, which can satisfy the requirements of passengers of different cultures (Western culture and Eastern culture). Even within the same culture, there may be differences, so the basic French breakfast and the English breakfast are completely different.

6.3.4 Situational, socio-economic, demographic or psychographic criteria

In addition, groups of passengers who share common characteristics can be determined on the basis of situational, socio-economic, demographic or psychographic criteria. Situational criteria encompass the following variables: sales channels (travel agents, online or phone), time/date of the flight, time of booking, location and access of origin and destinations airports, seat and ticket availability, ticket flexibility, loyalty/frequent flyer benefits, airport services and in-flight services. Personal characteristics of individual passengers are used in socio-economic and demographic segmentation, such as gender, nationality, religion, age, physical (dis)abilities, relationship status, income, first language, occupation, education/qualifications, whether passengers are travelling alone, in a group, in a family group or with babies and young children. Psychographic segmentation is related to the process of dividing passengers into groups based on their behaviour, travel motivation, values, attitudes, interests, personality and lifestyle criteria. This segmentation includes the following variables: trip motivation (the reason for travelling), destination, length of flight (short-haul or long-haul), total time away from home, travel class (economy, business or first class), travel experience: frequency of flying, cultural background of the passenger, airline preference (full service, low-cost or charter; global airline alliance, safety and service standard, brand value and reputation), membership in the airline's or airline alliance's loyalty programme, seat preference and environmental consideration (age of the aircraft and the airline's environmental record) (Wittmer and Hinnen 2017).

6.3.5 Business and non-business segments

As many airlines mainly focus on the basic air transport market segmentation—business and non-business passengers, it is interesting to see their shares in the air transport (globally) over the past decades (Table 6.1).

As can be seen from Table 6.1, decades ago business travellers were more dominant than non-business travellers. But nowadays, the situation has changed completely. The reason is the growth in the total leisure market, as well as growth caused by the shift in demand from another transport mode to air transport (Graham 2010). So, there now are much more leisure travellers, including tourists, those who travel to visit relatives and friends, etc. But one should bear in mind that business travellers are a very important category of passengers for an airline. Namely, business passengers are ready to pay a higher price for the air transport service.

Table 6.1. Business and non-business passenger shares over the past decades.

	Business segment	Non-business segment
1970s	60%	40%
1980s	50%	50%
1990s	35%	65%
The first decade of 21st century	20%	80%

Source: Data compiled from *Air Transport: Strategic Issues in Planning and Development* (Shearman 1992) and *Global Airlines: Competition in a Transnational Industry* (Hanlon 2007)

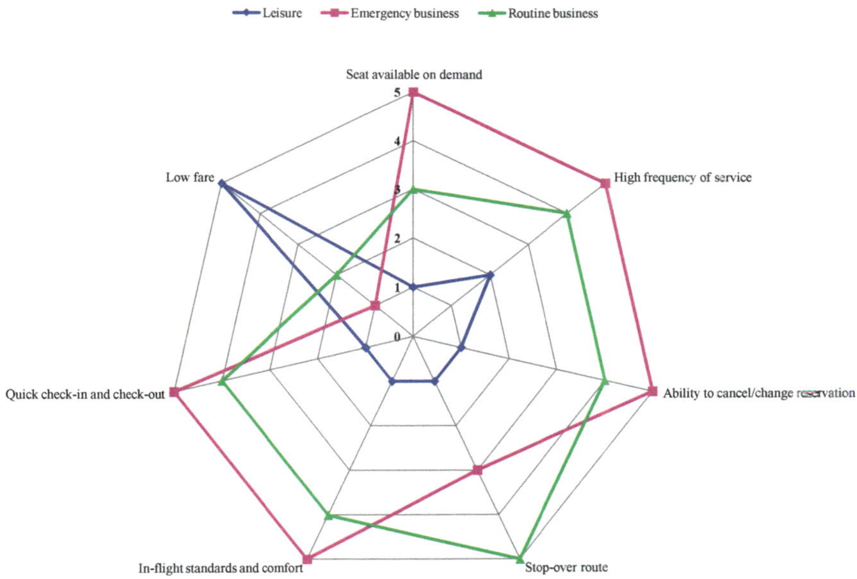

Figure 6.3. Passenger requirements by different segments. (data taken from Doganis 2010)

Different segments have different needs. Business passengers are mostly interested in high flight frequency, seats available on demand, ability to cancel or change their reservation, high level of in-flight comfort, quick check-in and check-out, etc. In contrast, the main requirement of non-business travellers is a low fare (Figure 6.3).

6.3.6 Identification of passenger segments

To identify passenger segments, airlines use well known quantitative (survey) and qualitative research methods (focus groups). The survey can be based on revealed and/or stated preference techniques. Revealed preference provides data on actual passengers' behaviour, while stated preference is based on hypothetical examples/questions which provide information about what potential passengers say (claim) they prefer. Analysis of data obtained by survey implies using basic and advanced statistical methods.

It is clear that the airline, by conducting research, most often surveys, is trying to collect information for different market segments that relate to the requirements of passengers, and their expectations regarding the air transport service. In this way, the airline's survey respondents are passengers who have already chosen this airline for their trip. It is very important and necessary for the management of the airline to carry out comprehensive research of all market segments, not only the users of the airline but all potential users in general. In other words, the survey sample must cover both passengers of the observed airline, and passengers of competing airlines in a given market, and other passengers who have chosen to use other competitive modes of transport. Thus, the management of the airline will have a full picture of all its market segments and based on that, create a business policy according to which it will create an appropriate product that will largely satisfy the market segments—target groups. These types of quantitative research are mostly outsourced to specialized consulting firms.

To emphasize once again, the goal of market segmentation is to gain a better insight into the market and to understand passengers' needs, which will lead to proper marketing decisions and designing appropriate air transport services. Depending on their business model, as well as their financial conditions, airlines implement different approaches to market segmentation, from simple to expensive, complex, and sophisticated passenger segmentation.

6.4 Airline performance measures

A key performance indicator (or KPI) is a metric that is one of the most important indicators of the current performance level of an individual, department, and/or company in achieving goals.[1] A key performance indicator (KPI) is a value used to monitor and measure how an airline operated in a given period. KPIs show different aspects of results. Thus, airline KPIs could be divided into five distinct areas (Table 6.2). Key performance areas (or KPA) encompass financial, customer satisfaction, airline operations, company resource efficiency, and the environmental

[1] https://www.optimizesmart.com/understanding-key-performance-indicators-kpis-just-like-that/

Table 6.2. Airlines KPAs and KPIs/metrics.

Key performance areas (KPA)				
Airline operations	**Financial**	**Customer**	**Airline resources**	**Environmental**
Key performance indicator (KPI) or metrics				
RPK ASK Load factor Number of passengers Number of flights Number of cancelled flights Number of delayed flights	RASK CASK Operating margin Operating profit per employee Revenue per employee Labour cost per operating revenues	Punctuality (on-time arrival) Mishandled baggage Oversells Consumer complaints	Aircraft utilization per day Number of employees per aircraft Number of passengers per employee ATK per employee	CO_2 emissions Nitrogen oxides (NO_x) emissions Climate index Noise Fuel consumption, airline operations Fuel consumption, ground operations

field. It is very important to understand the difference between an indicator and a metric. Metrics are basic and simple measures, such as the number of passengers or the number of flights. A metric is a quantifiable measure that is used to monitor and assess the current performance of a specific process. KPIs are more sophisticated, and mostly represent the ratio between two airlines' resulting outputs.[2] Many indicators can be taken from the annual reports of airlines. Some of them are presented in Table 6.2.

6.4.1 Airline operations area

Depending on the airline's capacity management, its well-designed flight schedule that synchronizes resources, pricing policy and offered fares, the airline will achieve appropriate results. How many passengers the airline will transport, how many flights it will perform, how many flights will be cancelled, and how many flights will be delayed, are the results of the operating business itself, as well as the airline's supply, including its matching with the demand. In addition, there are external factors that can affect the performance of air traffic (unexpected disruptions, recessions, pandemic diseases, unstable political situations, etc.). The well-known indicators and metrics related to airline operations are available seat kilometres, revenue passenger-kilometres, load factor, number of passengers, number of flights, number of cancelled flights, and number of delayed flights.

[2] https://professorannibal.com.br/2017/10/19/do-you-know-the-difference-between-metric-performance-indicator-index-unit-of-measure-and-target/

Airlines produce a product that can be expressed through the **available seat kilometres** (ASK) indicator. This is an indicator of available capacity as a result of supply (airline input). According to the definition (ICAO Glossary), this indicator is equal to the sum of the products obtained by multiplying the number of seats available for sale on each flight stage by the corresponding stage distance. A flight stage is the operation of an aircraft from take-off to its next landing. The unit of this indicator is *skm* (seat kilometres).

Revenue Passenger-Kilometres (RPK) is an indicator that represents the measure of the passenger traffic performance and it is the result of demand (airline output). RPK is calculated as the sum of the products obtained by multiplying the number of revenue passengers carried on each flight stage by the corresponding stage distance. Revenue passengers are passengers who paid for the air ticket, including passengers travelling under publicly available promotional offers or loyalty programmes, passengers travelling as compensation for denied boarding, passengers travelling on corporate discounts, and passengers travelling on preferential fares (government, seamen, military, youth, students, etc.). The RPK indicator excludes persons travelling free, persons travelling at a fare or discount available only to employees of air carriers or their agents, or only for travel on business for the carriers and infants who do not occupy a seat (definitions from ICAO Glossary). The unit of this indicator is *pkm* (passenger kilometres).

Load factor represents the utilization of capacity. In other words, this is an indicator of fleet productivity. It is calculated as the ratio between the airline output, RPK, and the airline input, ASK (LF = RPK/ASK). The load factor is a dimensionless measure, which means that by multiplying it with 100 this indicator can be expressed as a percentage. To evaluate the productivity of particular flights and to determine the level of commercial feasibility, airlines use the 'break-even load factor'. Airlines calculate the load factor at which they will break even. At a load factor lower than the break-even level, the airline will incur a loss, and above this level, it will record a profit. In other words, the break-even load factor is the load factor at which costs and revenues are equal. High or even very high load factors, in themselves, do not ensure profitability if yields are lower than costs (Doganis 2010).

The number of performed flights, number of cancelled flights, number of delayed flights, and number of carried passengers represent simple metrics and demand outputs that airlines monitor during different time periods (day, week, months, and year).

6.4.2 Financial area

This key performance area is related to airline economics and indicators based on revenue, cost and profit. The financial area generates the following KPIs: RASK (Revenue per Available Seat Kilometre), CASK (Cost per Available Seat Kilometre), operating margin, operating profit per employee, revenue per employee, labour cost as a percentage of revenues, etc.

RASK represents the revenue per unit of output produced, while **CASK** is the cost of producing each unit of output. RASK is calculated as the ratio between revenue and available seat kilometres. CASK is the ratio between cost and available seat kilometres. **Operating margin** is the ratio between operating profit and operating revenue, multiplied by 100. According to the definitions, operating profit is the total earnings from a company's core business operations, before deductions of interest and tax, while operating revenue is the revenue that an airline generates from its primary business activities (the transport service). **Operating profit per employee** is a metric calculated as the ratio of operating profit and the total number of full-time employees. This indicator shows how much profit each airline employee brings in a given period. To determine the efficiency and productivity of the average airline employee, **revenue per employee** is used. This is a rough measure calculated as the ratio between total revenue and the total number of full-time employees. Total revenue is obtained from ticket sales, cargo revenue, and ancillary sources (additional payments for food and drink on-board, extra baggage and priority seating). Labour costs represent one of the largest airline expenses, and are most often followed by fuel expenses. The result of dividing operating revenue with labour cost is the **labour cost per operating revenue** indicator, which shows how much money the airline spends on its employees in regard to the revenue from air transport service. Labour costs as a percentage of revenues are obtained by multiplying the ratio value by 100.

6.4.3 Customer satisfaction area

Customers are satisfied if there are no delays, no cancelled flights, no mishandled baggage, if the staff is friendly and professional, and if there are no oversells. Unfortunately, sometimes passengers find themselves in unexpected situations. To improve the level of service, airlines monitor indicators that affect customer satisfaction, such as punctuality of services, mishandled baggage, oversells, and consumer complaints.

Punctuality of services can be expressed through the ratio between the number of flights performed on time and the total number of flights carried out during a time period (day, month, or year). It is generally accepted that a flight is considered to be delayed if the delay is greater than 15 minutes. Therefore, if the delay is less than 15 minutes, this means that the flight was performed on time.

Mishandled baggage is delayed, damaged, lost, or stolen baggage. The indicator can be the ratio between the number of mishandled baggage and the total number of handled baggage during a given period.

Sometimes, to maximize profits, airlines **oversell** seats on certain flights. Based on historical data relating to no-show passengers on a particular route, airlines sell more tickets than available seats on an assigned aircraft. Thus, the boarding of passengers who had confirmed reservations was denied, because more passengers showed up than there were seats available on the plane. The indicator related to oversell could be the number of involuntarily denied boardings per 10,000 passengers in one month (Janić 2021). According to the report *Airline Quality Rating*,[3] **consumer complaints** include twelve attributes as criteria: flight problems, oversales, reservations, ticketing and boarding, fares, refunds, baggage, customer service, disability, advertising, discrimination, animals, and other. This indicator is used as consumer complaint ratio, calculated as the number of complaints per 100,000 passengers.

6.4.4 Airline resources area

Efficient resource utilisation is the goal of each airline. Airline resources are very expensive, especially aircraft and flight crews. Airline management tends to increase productivity, reduce staff numbers, introduce new technologies, and tackle rapid expansion (Shearman 1992). Airline resources as KPA include the following KPIs: aircraft utilisation per day, number of employees per aircraft, number of passengers per employee, ATK per employee, etc. **Aircraft utilisation per day** is calculated as the total block hours divided by the total number of days. Block hours of one flight are aircraft hours, from ramp to ramp, including flight hours. The total number of days is related to aircraft days assigned to the service in a given period. **The number of employees per aircraft, number of passengers per employee, and ATK per employee** are indicators that refer to employees' productivity. Available Tonne Kilometres (ATK) as a measure of airline's capacity, is expressed as the number of tonnes available (passengers, baggage, and cargo) multiplied by the flown distance.

[3] In the United States, domestic airlines collect data on passenger complaints. These data are published officially by the U.S. Department of Transportation.

6.4.5 Environmental area

The environmental impact of air transport has two aspects, global and local. The global aspect is related to climate change and global warming. The local aspect refers to noise disturbance, local air quality, water consumption, biodiversity and landscape effects, etc. During the flight, jet fuel emits greenhouse gasses, such as carbon dioxide (CO_2), nitrogen oxide (NO_x), and water vapour, so airlines contribute to air pollution and global warming (Budd 2010).

CO_2 emission can be expressed through relative or absolute values. Absolute emission of CO_2 is related to the total quantity of emissions, often measured in tonnes. **NO_x emission** is measured in tonnes, too. **Relative emissions of CO_2** refer to emissions per production unit and they can be expressed in grams of CO_2 per passenger kilometre or grams of CO_2 per tonne kilometre (gCO_2/pkm or gCO_2/tkm). The **climate index**[4] is calculated by taking the number of emissions of carbon dioxide and nitrogen oxides in relation to production. From the airlines' point of view, the **noise** level per aircraft movement must be considered. Noise emissions relate directly to the noise energy emitted by aircraft. Certificated noise levels and quota counts could be used as indicators. Total **fuel consumption** for both air operations and ground operations can be expressed in 1,000s tonnes or 1,000s litres. The indicator for fuel consumption, in the case of airline flights, could be the 'average fuel consumption of flights' and it can be expressed as kg per passenger kilometres (kg/pkm) or as litres per 100 pkm (l/100pkm).

6.5 Loyalty programmes and airline alliances

6.5.1 Loyalty programmes

In the period following the deregulation of the air transport market in the United States, many technical and legal constraints were removed, which enabled the airlines to define their services more freely, but they also had to do more marketing and to stand out due to increased competition. One of the results of the increased competition was the development of loyalty programs, known as the Frequent Flyer Program (FFP). The concept of the FFP was to reward high-frequency customers by giving them a free ticket after they reach a certain threshold of travel.

The first frequent flyer program was created in 1979 by Texas International Airlines, but this airline merged with Continental in 1982, and thus the program was terminated. The next FFP was launched by

[4] https://www.sasgroup.net/sustainability/sustainability-related-kpis/climate-index/

Western Airlines (Travel Bank program) in 1980, but it eventually became part of Delta's program when these two merged in 1987. In 1981 American Airlines' AAdvantage program was launched and it has remained active to this day. The initial purpose of the FFP introduction was to retain the most frequent passengers by rewarding them for their loyalty. Moreover, the main role of FFPs at the beginning 1980s was to distinguish a product of an airline from the products of its competitors and thus help the airline increase its market share. Today, almost all airlines offer loyalty programs, which significantly reduces this effect, if compared to the period when only some airlines had FFPs. Nevertheless, the researches indicate that FFPs continue to influence the stimulation of market demand (between 20 and 35 percent), as well as the competition through their FFPs' special offers or generous upgrades (Morrell 2007).

An FFP is a loyalty program that is designed for passengers, members of the program (mostly free of charge), to accumulate points (known as *frequent flyer miles*) corresponding to the distance flown with a particular airline. Frequent flyer miles represent a standard unit for measuring passenger's loyalty. Passengers collect frequent flyer miles every time they fly on a flight by the airline that owns the loyalty program or its associated partners. Once they have collected enough miles, passengers are eligible to exchange those miles for, e.g., a free ticket for a particular flight, an upgrade to a business or first class ticket, dedicated lounges at airports, or another reward depending on the airline's offer (Morrell 2007). Passengers can usually earn more miles for business and first class seats.

Over the years, FFPs became "the biggest and most successful innovative marketing program in the airline industry" and to be able to continue to develop, two preconditions are necessary:

- a deregulated and liberalized market, to create an appropriate market environment, and
- the existence of computer systems that provide appropriate infrastructure and support for the storage and management of large databases.

Creating a large database that contains historical data for its passengers allows an airline to assess the value of individual customers, and to run targeted marketing campaigns accordingly. This is why the purpose of customer tracking and database marketing is mandatory in developing a good FFP (Basso et al. 2009).

Regardless of the structure and complexity of FFPs, they are designed to increase the number of repeated purchases and to maintain loyalty in highly competitive markets, particularly among business travellers. A well designed and adequately implemented FFPs can enable airlines to automatically collect data, get to know their passengers, their buying habits and purchasing power, conduct customer database analyses and customer segmentation and offer services according to their needs. In this way, airlines offer their customers an additional reason to buy their services, and for loyalty, customers are rewarded with points that can be used later for various rewards. Also, loyalty programs have proven to be very good in attracting new passengers, preventing passengers from switching to the competition, differentiating the service from the service of competitors, and maintaining profitability (Tretheway and Oum 1992). Generally, the main idea of FFPs is to encourage people, particularly business travellers, not only to use the service of the same airline but to accumulate miles by travelling unnecessarily, more often, on longer flights and on higher tariff classes.

Most FFPs include a large number of partners, such as associated airlines, hotels, rent-a-car agencies, restaurants, shops, credit cards, phone companies, and other companies not anymore directly related to the travel sector. If the airline is a member of a global alliance, then its frequent travellers can accrue and redeem miles on all flights with the alliance partners. All the partners see the great advantage of offering loyalty programs to their clients, as well as great cost savings if they do so in association with airlines (Chin 2002). Thus, FFPs allows members to accumulate FFP miles not just on the airline's flights and flights of associated partners, but also when renting cars, staying at the hotels, using credit cards, etc.

Offering an FFP generates certain costs and revenues for the airline. The costs of loyalty programs can be divided into the following groups:

- Operating costs: employees, computer and communication equipment,
- Communication costs: promotional materials, program brochures, etc.
- Reward costs depend on whether a program member uses the reward ticket on a flight by the FFP owner, or on a flight by its partner.

If the passenger uses a free ticket on the flight by the airline that is also the owner of the loyalty program, these costs comprise (Tretheway and Oum 1992):

- Direct costs—in-flight service, additional fuel, reservation costs, etc.;
- Displacement costs—when the passenger who uses the reward ticket occupies a seat which would otherwise be occupied by a passenger who pays for the ticket;
- Diversion costs—when the passenger uses miles collected to buy a ticket instead of buying the ticket with money.

Direct costs always exist when the passenger uses the reward ticket, while displacement and diversion costs exist only if all seats are occupied, in the first case, and whether that passenger would have travelled even if he/she did not have the reward, in the second case. To prevent displacement and diversion costs many airlines have introduced certain restrictions when it is possible to use the reward as a free ticket. For example, free tickets cannot be used on flights during the holiday periods, such as Christmas and Easter, or it is predetermined for which flights the free tickets refer to. Since the airline has the cost of creating the FFP, as well as costs for its implementation and maintenance, it is estimated that the percentage of increase in ticket prices is in the range of 10–15, compared to the case if there is no loyalty program (Tretheway and Oum 1992).

If the passenger uses the free ticket on a partner airline's flight, the airline that owns the loyalty program pays a certain commission to that company, i.e., pays the reward for its passenger. The price that the airline pays in that case usually ranges between 1–2 cents per mile won.

Revenue that an airline can generate through a loyalty program comes from two sources:

- Membership fees paid by partner companies and
- Reduced advertising costs.

In order for an FFP to be attractive to passengers, the airline must meet certain requirements. First of all, the prizes must be achievable from passengers' point of view, i.e., the limit when they can win the wanted prize should not be set too high. For example, a loyalty program that offers rewards after winning 5,000 miles will be twice as attractive to travellers as a loyalty program that offers rewards after winning 10,000 miles. However, such a loyalty program will be more expensive and less efficient from the airline's point of view. Therefore, the airline must carefully consider its position in the market, the product it offers, and many other factors when making the decision to set limits for winning the offered prizes.

Creating an efficient FFP starts with a clear definition of the goals that the airline wants to achieve. Firstly, it is important to define the targeted group. The variety and value of rewards that will be offered through the loyalty program will vary between airlines, precisely due to the different characteristics and requirements of their targeted groups. Afterward, a well-designed FFP should help the airline keep its passengers by creating barriers to exit and switching to competitors. Moreover, it should prompt passengers to make additional purchases, provide insight into passengers' behaviour and preferences, and last but not least, generate a profit for the airline. FFPs can often be a lucrative source of income and profit (e.g., United Airlines went bankrupt in 2002 and the only profitable part of its business was the FFP) (Hanlon 2007).

The relative attractiveness of an airline's FFP can be affected by its program scheme, i.e., how travellers accumulate and redeem their rewarding miles. FFPs usually differ in how quickly it is possible to accumulate miles, depending on the tariff classes, type of flight, or even the FFP member status. Moreover, FFP partners such as airlines, hotels, restaurants, etc., can also be the main reason for passengers to choose one loyalty program over another. Finally, the conditions for using the miles, such as the validity period of the miles, the possibility of giving or buying miles, the possibility of using prizes on the flight provided for FFP awards, etc., are also the driving factors when choosing the right FFP (Chin 2002).

6.5.2 Airline alliances

An airline alliance can be defined as an arrangement between two or more airlines that agreed to cooperate on a substantial level in a certain time period. According to Cento (2009), the reasons for creating alliances can be divided into four groups:

- air-political, i.e., the airline has no traffic rights and is precluded from controlling a foreign carrier that has this right;
- economical, e.g., its costs or fleet (aircraft range and/or capacity) are unsuitable for that market;
- infrastructural, e.g., slots are not available; or
- financial-related, e.g., the airline has insufficient resources to develop new markets.

Airline alliances represent the most complex type of cooperation among airlines, defined by agreements that may include: code sharing; franchising, revenue and cost sharing; joint pricing; coordination of capacities; route and schedule planning; coordination of marketing, advertising, sales, and distribution networks; coordination of travel agents and other commissions; branding/co-branding; integration and development of information systems; information technologies and distribution channels; coordination of frequent flyer programmes; sharing of facilities and services at airports. The most common agreement between airlines is code-share, which is a commercial agreement between two or more airlines agreeing to offer the same flight under their own airline designator and flight number, while the flight is operated by one of them which is known as the operating airline. According to Holloway (2008), franchising of airlines involves a substantial airline with a strong brand, also known as the franchisor, licensing the use of the brand identity (e.g., aircraft livery, cabin design, crew uniforms, signage, logos, and brand names), the associated service concept, and aspects of the service delivery system to a franchisee (for a defined period of time). The franchisee pays the franchise for using the franchisor's brand, and to provide travellers the same experience as in the case when they would be travelling by the leading airline (the franchisor).

Alliances can be divided into marketing (commercial) and strategic alliances. In marketing alliances, partners follow their own goals and are independent, while in strategic alliances, partners work together to achieve common goals that are defined in advance. In strategic alliances, partners offer a common brand and a uniform service standard.

According to geographical coverage (Doganis 2001), alliances can be classified as:

- route-specific alliances that cover one route or a limited number of city pairs,
- regional alliances involving the joint coverage of routes in a particular region or country, or the franchise relationship of a larger airline and a smaller regional or feeder airline operating local traffic under the name and in accordance with the standards of the franchisor, and
- global, i.e., mega alliances.

Global alliances imply global coverage. They have a global network, which is the result of merging the networks of all member airlines, originating from different parts of the world. They provide a synergy effect that enables economies of scale, lower costs, and sustainable growth. Member airlines are the leading airlines in the world and medium-sized airlines from different parts of the world, thus their route networks create a global route network.

An alliance offers different possibilities to both airlines and passengers. The possibility of establishing traffic in markets that are very difficult to access or are inaccessible through connecting airlines from different countries initiated the formation of airline alliances (Vasigh et al. 2018).

Advantages for passengers are reflected in a greater choice, i.e., more destinations and flights are available as the network is expanded, travel times are shorter because schedules are coordinated, in the passenger loyalty program miles are collected faster since all members are involved in the joint program, better connectivity is provided (greater possibilities for connecting origin and destination) and lower prices are offered due to lower operating costs on a particular route. Advantages for airlines that are members of a global alliance are a better offer resulting from the connection of different markets, network optimization through resource utilization (flight frequency and passenger load factor) and traffic realization (schedule coordination), reduced competition as well as operating costs.

In addition to the above advantages for both passengers and airlines, negative effects can be observed, as well. Thus, higher prices can be offered on routes where there is no competition. Also, the frequency of some flights on a specific route may decrease due to network optimization, and the level of service will be reduced accordingly. Airlines that want to join an alliance have large initial costs due to the introduction of a new information system (used in the alliance). Moreover, airlines must announce 6–12 months in advance their willingness to join an alliance. It may also happen that airlines from the alliance are competitors on certain routes (Delta and Continental were competitors on transatlantic routes even though they were members of the SkyTeam alliance (Vasigh et al. 2018)).

Leading global alliances on the airline market are Star Alliance, Oneworld, and SkyTeam. Each of the mentioned alliances will be briefly presented below.

Star Alliance was founded in May 1997 and it was the first global alliance, and it is still the largest one. Its founders were Air Canada, Lufthansa, Scandinavian Airline System (SAS), Thai Airways International, and United Airlines. The alliance is headquartered in Frankfurt, Germany.

The goal of Star Alliance is to improve the quality of services provided, which is achieved through a large number of offered destinations and opportunities for easier, faster (joint route network, check-in, and ticket sales) and more comfortable travel (members are located at the same airport terminal, they build facilities at airports together, provide access to lounges, coordination of flight schedules and setup connection teams to make the transfer faster). All airlines that are members of Star Alliance have a uniform standard, as well as the level of service. In addition, all airlines that are owned by the member airlines and have the status of associated airlines and provide passengers with all the benefits as the member airlines. These benefits include a global route network, reduced connection times, simpler and faster transfers, unique passenger loyalty programs (members of any loyalty program collect miles on flights of all member airlines), access to airport lounges, connection centres (flights are tracked, and in case of delay, a transfer is arranged before landing). The alliance strives to locate all member airlines at the same terminal, especially at large airports, to significantly reduce connection times.

Star Alliance airlines are Aegean Airlines, Air Canada, Air China, Air India, Air New Zealand, All Nippon Airways (ANA), Asiana Airlines, Austrian Airlines, Avianca, Brussels Airlines, Copa Airlines, Croatia Airlines, Egyptair, Ethiopian Airlines, EVA Air, LOT Polish Airlines, Lufthansa, SAS Scandinavian Airlines, Shenzhen Airlines, Singapore Airlines, South African Airways, Swiss International Air Lines, TAP, Thai Airways, Turkish Airlines and United.

Oneworld Alliance was founded in September 1998 by American Airlines, British Airways, Cathay Pacific, and Qantas. Its headquarters are in Vancouver, Canada.

The goal of Oneworld is to provide passengers with as comfortable travel as possible, to offer them as many rewards as possible within its passenger loyalty programs, and to offer more new travel products that will find their customers among business and leisure travellers. In the Oneworld alliance, there are affiliated airlines (performing regional traffic together with one of the members of the alliance) that provide the same level of service as the member airlines. Passengers are provided with more support and assistance by employees (both on the ground and in the air), there are higher rewards for passengers who are members of the passenger loyalty program, transfer is easier (employees of all member airlines are available to passengers, staff and desks at airports are clearly marked, check-in facilities are offered), a large number of destinations are offered, access to airport lounges is provided (depending on the status in the loyalty program).

Oneworld alliance airlines are Alaska, American Airlines, British Airways, Cathay Pacific, Finnair, Iberia, Japan Airlines, Malaysia Airlines, Qantas, Qatar Airways, Royal Air Maroc, Royal Jordanian, S7 Airlines, and SriLankan Airlines.

SkyTeam is an alliance founded in 2000 by the following airlines: Aeromexico, Air France, Delta Air Lines, and Korean Air. Its headquarters are in Amsterdam, the Netherlands.

SkyTeam is dedicated to passengers, striving to provide a more comfortable and easier journey, as well as to keep passengers informed during the travel. The member airlines allow passengers to access the global air traffic network through their hub-and-spoke network.

When it comes to benefits that the member airlines have, strengthening their own brand through the brand of the alliance and other members is one of the benefits worth mentioning. Through cooperation, defined by various agreements, the members exchange valuable knowledge and their practical application, especially in the field of security, services provided, and operational efficiency. For airlines wishing to become members of the alliance, a desired standard relating to the level of service must be met. When an airline joins the alliance, it expands its route network through partnerships and various agreements within the alliance. To achieve cost savings, the member airlines maximize airport capacity utilization by sharing the working space and terminals at airports. Their check-in counters are located in one place, staff is combined for better efficiency, ground handling costs are reduced, and the use of airport lounges is optimized.

Table 6.3. Comparative overview of leading global alliances (2019).

Global alliance	Star	Oneworld	SkyTeam
Member airlines	26	14	19
Countries served	193	180	175
Airports served	1300	1100	1050
Passengers (mil)	757	550	630
Lounges at the airports	1000	650	750
Fleet	4657	3439	3054

Data source: websites of the alliances

SkyTeam member airlines are Aeroflot, Aerolineas Argentinas, Aeromexico, Air Europa, Air France, Alitalia, China Airlines, China Eastern, Czech Airlines, Delta, Garuda Indonesia, Kenya Airways, KLM, Korean Air, Middle East Airlines (MEA), Saudia, Tarom, Vietnam Airlines and XiamenAir.

Table 6.3 provides a comparative overview of the leading global alliances.

6.6 Airline fleet planning

One of the main goals of an airline is to match its capacity and demand as closely as possible. However, at the same time, the airline needs to consider both its own and passengers' interest; it needs to make a profit and meet the expectations of passengers. Moreover, the airline needs to consider prevalent market conditions, and take appropriate actions to increase its profitability and to reduce its costs. Therefore, it is necessary to analyse the passenger demand in the past, as well as the influencing factors, to be able to make an adequate forecast. Adequate, appropriate forecasts for the airline also mean a high probability of successful business, due to the fact that all planning activities in the airline (network development, flight frequency determination, flight schedule design, fleet development, pricing policy) are based on the demand forecast.

During the process of airline fleet development planning, as already mentioned, passenger demand should be taken as the starting point. Based on past and current data, future transport demand should be determined. For a known demand, a route network should be designed and flight frequencies should be determined, and finally, based on these data, fleet planning can be approached.

The airline fleet planning process, i.e., determining the necessary capacity for the appropriate demand and market, is a very important task. The airline plans to develop its fleet depending on its business policy, the route network, and the markets it serves. Fleet planning in an airline involves great complexity, as well as great uncertainty. To have a satisfactory fleet planning result, it is necessary that engineers and experts from different fields (economics, management, law, etc.) participate in the planning process. Involving people with different educational backgrounds in the process of fleet planning enables the airline to see the effects of decision making from different perspectives. Decision makers and planners also need to be able to appropriately assess what will happen in the future, taking into account relevant factors.

To keep its market position, an airline needs to manage its fleet, i.e., it should permanently monitor the fleet and decide how many aircraft and when would be acquired (bought or leased) and retired. The existence of a large number of different aircraft types, which would have a certain mission and purpose in the airline services and which could operate markets that have different air travel demands and characteristics, emphasizes the complexity of the fleet planning and aircraft selection problem. Aircraft should be chosen to be used in the future while meeting air travel demand in given market conditions (price and competition) (Dožić and Kalić 2018).

From an airline's perspective, "fleet planning is the process by which an airline acquires and manages appropriate aircraft capacity to serve anticipated markets over a variety of defined periods of time with a view to maximising corporate wealth" (Clark 2007).

This definition should be explained in more detail. Firstly, planning is a continuous process, not a one-time decision making. After the decision regarding the fleet has been made, the planning process does not end, on the contrary, it continues. When it comes to aircraft acquisition, it usually refers to buying or leasing an aircraft. The decision to buy or lease is made based on factors such as finances, aircraft availability and delivery time, airline-specific requirements, etc. Fleet management means using the aircraft in an appropriate way, but also making a decision on its replacement, sale, change of purpose (sometimes passenger planes are converted into cargo versions), or leasing. The capacity of an aircraft describes its size. Depending on the frequency of the observed market, an airline makes decisions related to the size of the aircraft that corresponds to the needs and future plans. Decisions regarding planned (future) markets imply that something can be changed in the current market. Therefore, the acquired aircraft should correspond to both, the current and the future market. It is also important to note that an aircraft is a resource that is used for a longer period of time. For that reason, it is necessary to consider short-term and long-term plans, to make appropriate decisions. Namely, in the short term, one type of aircraft can meet the needs of the airline, while for the long-term period, such a decision may be unacceptable. Given the large investments required in the fleet acquisition, it is evident that the mission of aircraft is not to provide operations for a short-term period. Therefore, all aspects should be considered and the decision should be made so that the purchased aircraft can be used for a longer period of time. Finally, the needs of the airline should be harmonized with its capabilities, i.e., with the funds available for investment, and with profitability. The profitability of the airline affects the decision regarding the fleet acquisition, and vice versa, the fleet has an impact on profitability (if the fleet is not appropriate, business results will not be satisfactory).

According to the definition of fleet planning, it can be seen that the process involves making certain decisions that, as mentioned earlier, require very large investments. This is the reason why it is very important to know who the decision makers are, as well as who manages the fleet. Historically, in the beginning, the decision makers were pilots (Clark 2007). Their opinion was crucial when buying aircraft. Afterwards, the role of decision makers was given to engineers. Planning departments in airlines consisted of engineers mostly. During the process of fleet planning and making decisions related to the fleet, engineers evaluated aircraft available on the market, analysed take-off and en-route performance of the aircraft, checked the compatibility with the airports to be flown, system reliability, as well as maintenance requirements. This approach led to a rational and technically justified decision (Clark 2007).

After the air transport deregulation, the circumstances in the aviation market have changed. The airline business became market-oriented instead of regulated by governments. Therefore, the approach to fleet planning has also changed. Engineers, who were the decision makers, were replaced by experts in the field of marketing, as well as experts in finance. What has not changed and what is always set as an imperative is that security and safety need to be at the required level. As the aviation market is subject to various influences, and therefore is very volatile, airlines are required to adapt to new market conditions to survive in a strongly competitive market. Thus, the airline business started to be profit-oriented, rather than technically oriented. Moreover, there were teams made and involved in the decision-making process so as to determine route structures, fare policies, and brands, and to determine investment levels, sources of funds, and whether any money can be spent at all (Clark 2007). Due to these changes, engineers became support in the decision-making, as well as in the fleet planning process instead of being the actual decision makers. It is important to note that engineers' opinion is still of fundamental importance in all processes. Nowadays, planners in airlines have the role of project managers. They select an appropriate team and manage it, so that they give each of the team members the opportunity to participate in the final decision-making, giving the maximum contribution in their own field of expertise.

It is worth noting that some differences in fleet planning approaches can be observed depending on the airline type. For example, traditional and cargo airlines are advocates of the traditional (engineering) approach to fleet planning. Their goal is to harmonize supply and demand in the best possible way, taking into account all technical aspects. On the other hand, low-cost, charter, and leasing companies consider the fleet as a means to achieve set business goals.

6.6.1 Fleet characteristics

There are two basic features of a fleet and they are the fleet structure and the fleet size. The fleet structure represents the number of different aircraft types in the fleet. Fleet size represents the number of aircraft by different types and the total number of aircraft in the airline's fleet. Speaking of fleet structure, it should be mentioned that it could be homogenous (uniform or single fleet) consisting of only one type of aircraft, and heterogeneous (multi fleet) consisting of more than one aircraft type in the fleet. In the case of a homogenous fleet, maintenance, training, and labour costs are lower in comparison to a heterogeneous fleet. From the other point of view, different aircraft types in the fleet allow airlines to match demand and supply more closely, make high passenger load factors and improve their profitability. Therefore, it is very important to plan the fleet structure according to the airline's needs (Dožić and Kalić 2018).

Another important issue is the fleet size, which should be planned in accordance with the airlines' needs. Otherwise, the fleet size could affect the airline's business performance. Namely, the number of aircraft larger than required, for an airline, means lower fleet utilization, which further induces increase of costs (Figure 6.4).

On the other hand, less aircraft than required, for an airline, means spill of the demand, i.e., reallocation of passengers to competitor's flights and missing out on the opportunity to improve the profitability. Fleet size also needs to be determined according to the airline's requirements. Moreover, both fleet structure and fleet size must be determined properly

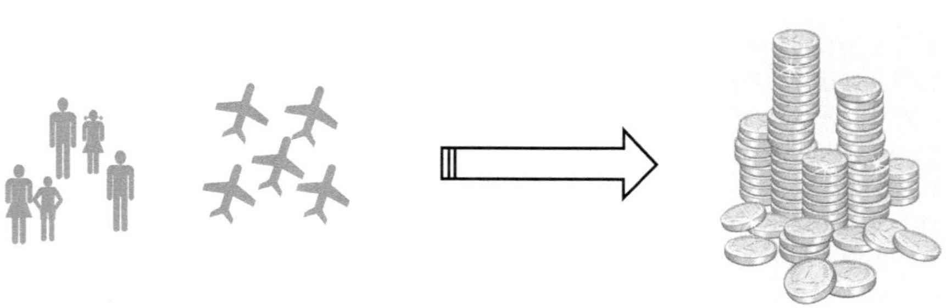

Figure 6.4. The fleet size is greater than the actual needs of the airline.

to enable the airline to realize the planned schedule and generate a profit (Figure 6.5) (Dožić and Kalić 2018).

Indicators of aircraft utilization (such as passenger load factor, annual flight hours, etc.) show how this resource is used. Passenger load factor can be defined as the ratio of the number of revenue passengers (passengers that paid for air tickets) and the number of available (offered) seats in the aircraft. Greater load factor and greater annual flight hours are closely connected to greater profitability, as well. This means that these indicators have a direct impact on revenues and costs, which further influence the airline's business performance. Whereas fleet structure, fleet size, and fleet utilization affect the airline's profitability, it is very important that the fleet corresponds to market requirements.

Bearing in mind that acquisition of a new aircraft requires a huge investment,[5] it is evident that a small savings of a few percent is not negligible for an airline. To achieve even such small savings, adoption of an appropriate methodological approach is essential (Dožić and Kalić 2015). Although costs of acquiring an aircraft represent large capital expenditures for an airline, these costs are very often lowered by attractive aircraft leasing options.

In addition to the fleet size and the fleet structure, another important feature of the fleet is its ownership structure. Airlines in their fleet may have aircraft which are owned, as well as aircraft used on the basis of lease agreements. Thus, the fleet ownership structure of an airline indicates the share of owned and leased aircraft in the airlines' fleet. It should be noted that there are airlines that have wholly owned fleets (e.g., Iran Air, Tunisair), as well as those that lease the entire fleet (e.g., Air Malta, Ural Airlines). The owned/leased aircraft ratio should be determined in a way that enables the airline to increase profitability and reduce the costs and it should be in accordance with the financial capabilities of the airline. Regarding the fleet ownership structure (Dožić and Krnić 2016), found

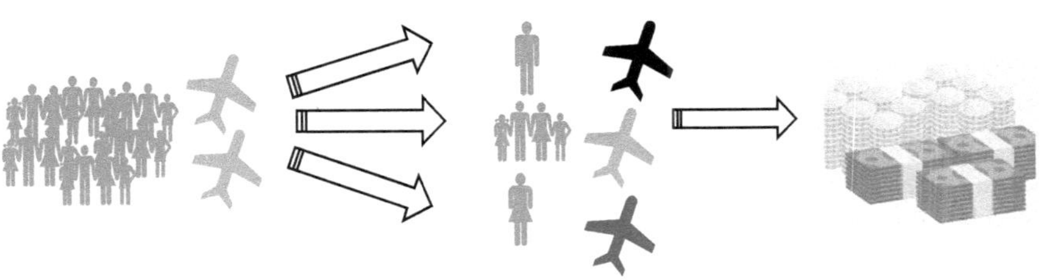

Figure 6.5. The fleet size is smaller than the actual needs of the airline.

[5] According to Airbus price list for 2018, the average price for an aircraft produced by Airbus, expressed in millions of $ is in the range of 81–445.6; according to the current Boeing price list (June 2021), the average price for aircraft produced by Boeing, expressed in millions of $, is in the range of 89.1–442.2.

that the percentage of leased aircraft is inversely proportional to the fleet size. Namely, a small number of aircraft in an airline fleet (less than 50) implies a higher percentage of leased aircraft (percentage of leased fleet is higher than 50), while a larger number of aircraft in an airline fleet (more than 100) implies a smaller percentage of leased aircraft (on average, approximately a third of the fleet is leased).

6.6.2 *Levels of fleet planning*

The development of an airline fleet can be planned at different levels: at the strategic, tactical, and operational levels.

Fleet planning at the strategic level takes into account the long-term business strategy of the airline, economic development, legal regulations, but also the policy of the airport where the airline will operate and aircraft manufacturers with their products which are offered and which will be offered on the market. Technical, operational, and economic characteristics of all commercial aircraft in use and aircraft in development that meet the requirements of the market where the airline operates should be considered. Decisions made at the strategic level are long-term decisions.

Tactical planning refers to the decision to acquire an aircraft based on fleet commonality regarding the existing fleet, the aircraft price, and the route network (hub-and-spoke or point-to-point) (Wei and Hansen 2007). These decisions can be considered short-term decisions.

Operational planning usually refers to the fleet assignment. More precisely, it means that decisions should be made to assign flights to each aircraft for each day of the week (decisions related to covering each leg by an aircraft for each day of the week). Operational decisions are short-term decisions.

6.6.3 *Aircraft type selection*

Traditionally, fleet planning has focused on aircraft acquisition, which involves aircraft type selection. Many large airlines now take a broader view, with fleet management encompassing aircraft acquisition and financing, tactical fleet management, asset value maintenance, and aircraft trading (Holloway 2008). Firstly, it should be decided which type of aircraft should be bought and how to finance it. When the aircraft becomes a part of the fleet, it should be managed and utilized properly, to maintain its value. Finally, it should be decided whether to change the mission of the aircraft or its purpose on the market, depending on the market conditions.

According to Holloway (2008), there are two main reasons for aircraft acquisition. An aircraft should be replaced because of inappropriate characteristics (high operating costs, unacceptable noise or emissions, limited remaining structural life, inadequate passenger appeal, type rationalization, or an ongoing fleet rollover policy intended to maintain a low average fleet age), and more effective aircraft which will have the same mission should be placed in the fleet. Capacity growth could be growth within the existing network driven by the increase of demand. Moreover, growth could also be induced by introducing a new service.

When an airline decides that there is a need for aircraft acquisition, the question which arises is: which aircraft is needed? To determine the most suitable aircraft, planners very often have to balance multiple, usually conflicting criteria. Interests of both the airline and passengers must be considered, as well as operational requirements. Aircraft should be evaluated, and the following issues should be considered (Holloway 2008):

- Who is the decision-maker? Experts with different educational backgrounds have different points of view (engineers focus on performance and maintenance, marketing specialists focus on product design, finance staff focus on operating costs and types of finance, and sometimes political factors come into play, as well).
- Collation of airline-specific data includes network data, airport data, current fleet, etc.
- Different analysis should be prepared, for example, marketing, technical, performance, cost, powerplant analysis.
- Avionics, options and customization, customer support analysis, the fleet commonality issue, and some other issues (delivery time, possibility of financial support, single/multi-fleet) should be considered.
- The whole process ends with the pricing and the economic analysis, and contracting.

Whereas aircraft type selection can affect the airline's profitability, it is necessary to make the right decision at the right time following a defined procedure that involves all relevant aspects and factors. Hence, the aircraft's purpose (cargo, training, or commercial passenger aircraft) and mission on the market (for charter, low-cost, regional, or full service airlines) should be examined and the aircraft should be evaluated accordingly (Dožić and Kalić 2018).

Based on different researches and considering criteria/sub-criteria presented in the literature and used in the process of aircraft type selection (Dožić et al. 2018), propose five aspects that could encompass all criteria. These aspects are:

- Technical/Technological aspects that comprise criteria related to aircraft and airport characteristics, as well as aircraft performances which could be found in manufacturers' manuals.

- Economic aspects, which are very often used and they include all criteria that are measured in monetary units, and which are mostly related to different categories of costs.
- Passengers' aspects refer to the quality of service and could comprise of various criteria important for passengers (safety, baggage, comfort, etc.). Some of them are provided by the manufacturers (for example, available space in the aircraft), while others (related to the subjective judgment of passengers) could be collected from market surveys.
- Environmental aspects that are becoming an unavoidable issue in recent years due to the fact that humanity faces the problem of pollution.
- Other aspects refer to criteria that could not be assigned to the previous four aspects, and they are usually measured qualitatively and may include various criteria related to delivery time, fleet commonality, payment conditions, robustness, etc.

It is noticeable that Technical/Technological and Economic aspects are dominant in the literature.

Based on aircraft evaluation, the airline can decide that an aircraft should be acquired, and also which one is the most acceptable. The next step is to consider the finances. It is necessary to separate decisions related to investment and financing (Holloway 2008). Investment is related to a decision on which aircraft should be acquired and at what cost, while financing refers to how the acquisition should be financed. Aircraft acquisition can be financed from the airline's own internal funds, or mixed internal funds and debt, and using leasing options. An aircraft lease can be defined as a contract whereby the owner of an aircraft (the lessor) grants to another party (the lessee) the exclusive right to the use of the aircraft for an agreed period, in return for the periodic payment of rent (the fee) (Morrell 2013).

As the air travel demand is directly affected by permanently varying conditions in the airline market, airlines need to respond to these changes and offer corresponding capacity as promptly as possible; therefore, they need a flexible fleet. With flexibility being an important issue, aircraft lease appears to be a suitable option for airlines (Dožić and Krnić 2016), due to an easier, faster, and usually cheaper way to provide appropriate aircraft.

6.7 Airline network planning

6.7.1 Airline network structure

The airline network design process is a key part of the airline's strategy, due to its very high impact on generated costs and revenues. The network design problem basically involves all decisions related to which markets to serve, how to design routes between the markets, what the consequences of introducing/eliminating certain routes are, what the consequences of changing traffic flow in the network are, etc. These decisions are primarily related to finding the right balance between serving different market segments, on the one hand, and meeting the economic interests of the airline, on the other hand (e.g., offering high-frequency non-stop flights is significantly more expensive than offering lower frequencies using larger aircraft).

A route connects two markets, and when the same aircraft (under the same flight number) is flying along the whole route this route is called (Holloway 2008):

- a non-stop route—there is only one segment between the origin and destination airports (e.g., Paris-Munich);
- a direct route—there are two or more stops between the origin and destination airports, (e.g., Belgrade-Larnaca-Dubai, is a direct flight performed by one aircraft under one flight number).

If the same flight number is used on a route, and the aircraft is changed on one of the intermediate stops, it is considered that these are two connected routes. Both routes can be served by the same airline, but in practice, it is often the case that the airline uses the flight number of the airline with which it has a code-share agreement on one of the routes. Another case is when two routes are served by different airlines that have another form of interline agreement.

Airlines use all of these options to serve their markets, and the particular choice mostly depends on the characteristics of demand segments present in that market and airline strategy. Each of these options will represent different services for passengers, and accordingly, different opportunities for airlines to generate revenue. Last but not least, when designing their networks, airlines take into account the impact on competition.

The most dominant network structures developed by airlines are point-to-point (PP) and hub-and-spoke (HS). Network structure in the airline industry is closely related to the business model of an airline. According to the current practice, the characteristics of a PP network are more suitable for the business model of low-cost carriers, while the HS network is more suitable for the business model of full service airlines. Of course, real world airline networks are often a mixture between these two different forms, with predominantly HS characteristics or predominantly PP characteristics.

6.7.2 Point-to-point network structure

In a point-to-point network, all airports are served directly with non-stop flights. There are two types of point-to-point network structure: the linear network and the grid network. They are mostly used by low-cost carriers, charters, and those national airlines that do not have the capacity to develop an HS network due to limitations at their base airports.

The linear structure of the network was a characteristic of European airlines in the period before the deregulation. In the linear structure, the aircraft starts from its base airport (which is not a hub) and makes several intermediate stops on the way to its final destination. Stops are performed to refuel or to board passengers along the route (Figure 6.6). Although there are airlines around the world that still use this type of route network, many airlines have abandoned the linear structure, due to high costs and inefficiency. Costs tend to be high, as they include airport costs at multiple airports on a single route, while revenues are often very small, due to the low frequency of service on these routes.

The grid network mainly includes routes of short and medium length (up to 3,000 NM), directly serving airport pairs (Figure 6.7). In this structure, the aircraft starts from its base airport and makes several intermediate stops, connecting these cities with each other, and ends its rotation at the airport from which the flight started. All passengers in a PP network board at the flight origin and disembark at the destination. The flight schedule is not coordinated at larger airports to offer connecting flights for passengers and baggage, so there is no responsibility for missed connections.

The PP network structure provides a high level of connectivity as well as low operating costs, but it is also characterized by the low quality of service in terms of flight frequency and capacity on the routes. When planning a PP network, the target group for airlines are local passengers whose origin and destination of the trip are at the departure or destination

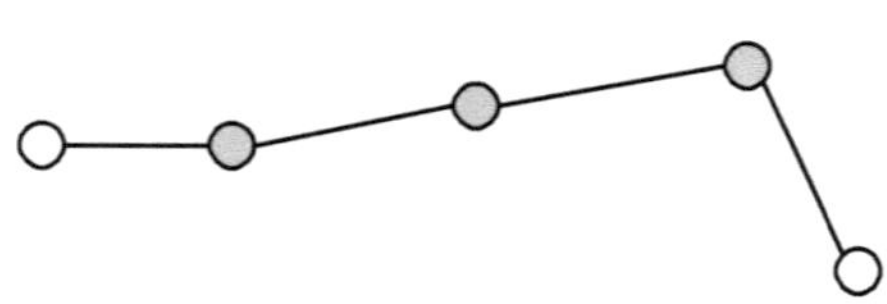

Figure 6.6. Linear network structure.

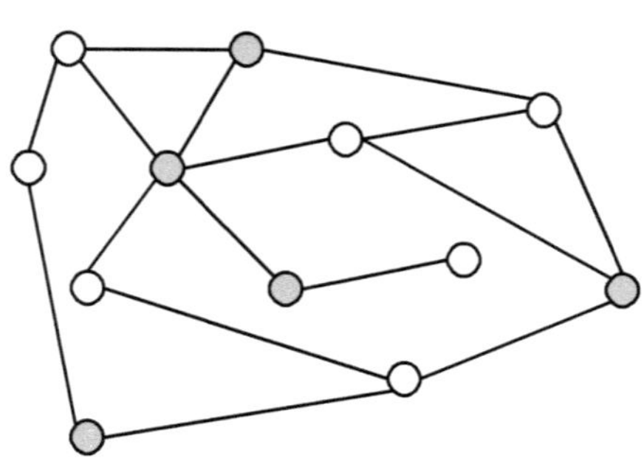

Figure 6.7. Grid network structure.

airport of the route, and not transfer passengers. Generally, a non-stop flight between origin and destination is the least expensive way to serve markets where demand is sufficient to support larger aircraft.

6.7.3 Hub-and-spoke network structure

An HS network consists of a hub, which is a central airport, and a large number of lines (spokes) that extend radially relative to the hub, to many surrounding airports (Figure 6.8). By using the HS system, an airline can easily provide air transport service to a wide geographic area and many destinations, directly or indirectly. In the HS network structure, the number of potential connections at the hub airport grows exponentially with the increase in the number of markets served from the given hub. This makes the hub, as well as the airlines that use this hub, very attractive from passengers' point of view.

The HS system became a dominant choice for full service airlines in the period after the deregulation for a variety of reasons. In addition to high connectivity, an HS network provides potentially higher revenues, higher operating efficiency, and less aircraft required to service the network compared to a PP network. This structure is suitable for many airport pairs that do not have enough demand to justify direct flights with a higher frequency, so passengers can use connected flights to reach these airports. By directing passengers to the hub, the airline achieves a significantly higher traffic density on each route compared to a PP network. The higher traffic density allows the airline to offer a higher frequency or to use larger aircraft on these routes, which additionally attracts new passengers.

HS networks are also used as barriers to the entry of new airlines into the market due to their ability to enlarge the scope of network coverage, (Holloway 2008). At the same time, by concentrating its flights at the hub, an airline is limiting competition at that hub because of the airport's capacity constraints. This limitation brings a demand advantage and substantial market power for airlines in their own hubs compared with competitors that operate routes outside the hub (Borenstein 1991).

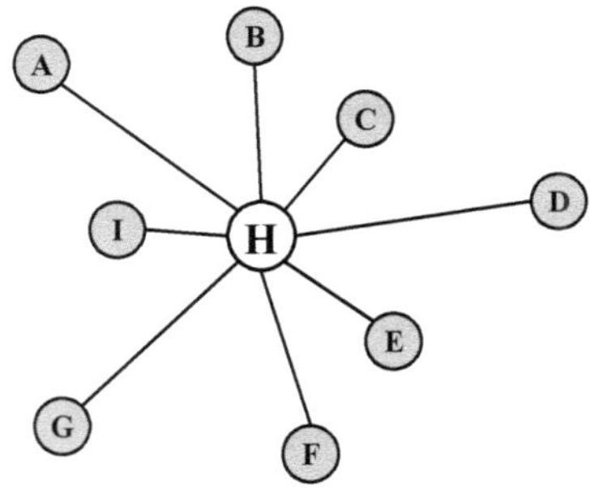

Figure 6.8. Hub-and-spoke network.

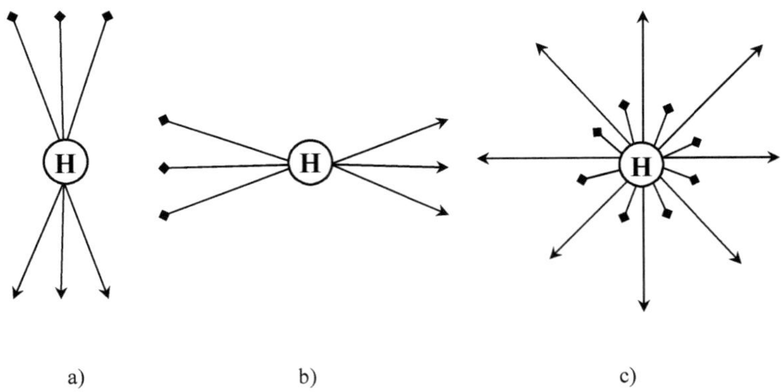

Figure 6.9. (a) Hourglass hub north-south; (b) Hourglass hub east-west; (c) Hinterland hub.

There are two types of hubs in terms of connections they offer: hourglass and hinterland (Figure 6.9). An hourglass or a directional hub (Figure 6.9a and Figure 6.9b) offers flights that are operated from one region through the hub to the points that are in a completely opposite direction with the intention of minimizing unnecessary detours. It usually connects the 'long-haul to long-haul' market or 'regional to regional' market and is often organized to connect north-south or east-west. Flights through a directional hub are usually operated with the same type of aircraft (passengers may switch to another aircraft, but not necessarily). Examples of hourglass hubs in Europe are Copenhagen airport (Scandinavia–Europe) and Vienna airport (East-West), in the US: Chicago and Dallas airports, and in the Middle East: Dubai airport.

A hinterland or a multi-directional hub (Figure 6.9c) is mainly organized in that way to link short-haul flights to long-haul flights. Airlines develop hinterland hubs to feed their long-haul flights with passengers from domestic and regional routes. Due to a significantly large difference between flight duration on short-haul and long-haul routes, passengers do not take much into account the problem with detours when making the connections. Flights through a hinterland hub often require a change of the aircraft type (from a regional aircraft passengers switch to a long-haul aircraft, and vice versa). There are several examples of hinterland hubs in Europe, such as Zurich and Amsterdam airports.

There are different forms of HS networks when it comes to the number of hubs and they are usually divided into single hub networks and multi-hub networks. In a network with one hub, all airports are connected to a single airport, which is the hub, chosen because of its central geographical location. This group also includes the typical HS system shown in Figure 6.10, where besides the routes directly connected to the hub, there

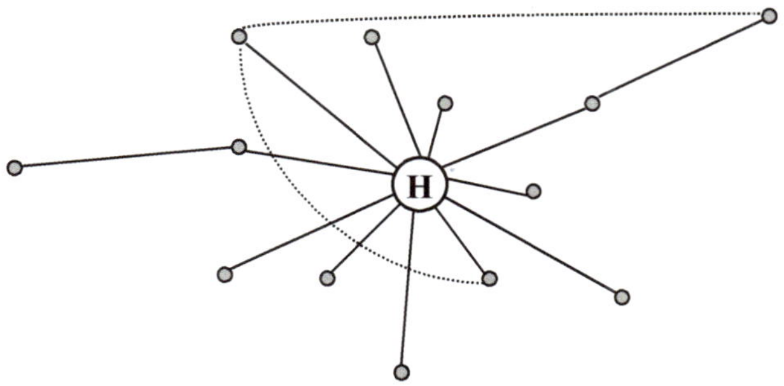

Figure 6.10. Typical HS network.

are direct routes between other airports in the network if the demand is sufficient to justify the operations (shown by the dashed line in Figure 6.10).

A multi-hub network is used to serve different regions, some of which include several primary and secondary hubs, Figure 6.11. In a multi-hub system, all airports are connected to at least one of the hubs, while the hubs are connected by non-stop flights (American Airlines has 5 developed hubs in different regions across the United States: Dallas/Fort Worth International Airport, John F. Kennedy Airport (New York), Los Angeles International Airport, Miami International Airport and O'Hare International Airport (Chicago)).

Multi-hub networks are widely used among global alliances, where the hubs of partner carriers on different continents are connected. Airlines that use multiple hubs in their networks face additional business decisions that need to be made: with how many hubs to connect secondary and tertiary airports, how to determine good connections between the hubs (flight frequency and capacity), as well as network flows (especially in places where there are more options).

For an airline operating at a hub airport, the most important issue is how to design a flight schedule to achieve the maximum number of connections, taking care to make those connections within a reasonable time frame. For that purpose, the airline operates a wave-system structure to maximize the number of connecting opportunities. In a wave-system structure, flights are organized in so-called flight waves, incoming and outgoing, structured so that all incoming flights connect to all outgoing flights. Two consecutive waves (the first is incoming and the second is outgoing) form a wave-system structure or a connection complex. Each wave-system is determined by a time frame, which represents the period from the moment of the first arrival at the hub in the incoming wave to the moment of the last departure in the (next) outgoing wave. The time frame

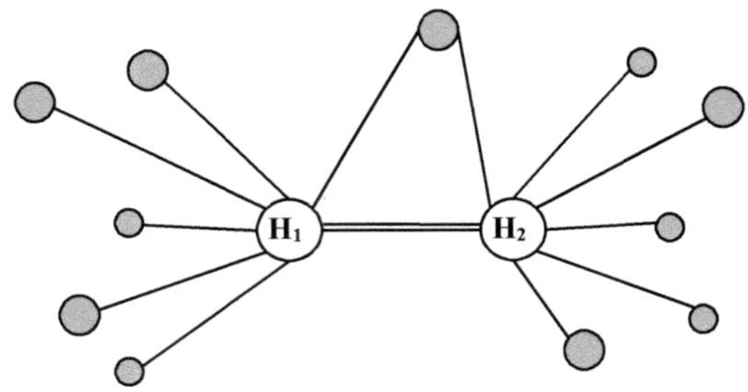

Figure 6.11. Multi-hub network.

defines the time for which passengers can make their connections in each wave-system.

6.8 Airline scheduling

Flight schedule represents one of the primary products of an airline and it is the leading factor when passengers choose an airline. It also represents a flight plan or schedule that contains the following elements:

- Day of operation,
- Departure time,
- Departure airport,
- Arrival time,
- Arrival airport,
- Number of intermediate stops (with appropriate arrival/departure airports and arrival/departure times),
- Aircraft type,
- Flight number.

Moreover, a flight schedule is an instrument to match the available resources to the given demand. An airline's flight schedule is the main product offered to the customers and it is also the main generator of the airline's revenues and costs. The operational success of the airline and its quality of service depends on the flight schedule, which is designed to fulfil passenger demand, utilise available resources, and satisfy different operational requirements. The greater the time buffer between two flights, the greater the probability of servicing the flight schedule according to the plan. However, at the same time, this means that the airline's resources are less employed and its income is reduced accordingly. The flight schedule design process aims to harmonize these two conflicting criteria (punctuality and utilization), to secure a reliable timetable and employ resources efficiently.

During flight schedule designing, different factors should be considered, as well as the interests of both passengers and airlines. Therefore, scheduling is one of the most complex and challenging tasks, as it implies searching for the balance between adequate service and economic strength for the company.

6.8.1 Airline schedule development

When it comes to scheduling, Holloway (2008) listed certain steps that should be followed during the process of airline schedule development. According to him, this process begins with the **choice of markets** to be served, based on demand forecast. To select the markets, it is necessary to conduct wider market research, as well as to take into account the corporate strategy of the airline, i.e., to choose the market in accordance with the defined goals. For some routes, it is necessary to obtain appropriate permits and licenses to be able to operate them. These elements are the key elements for route network planning. All costs related to the fleet, aircraft equipment, labour, and traffic, depending on both short term and long term, depending on the selected market and the defined route network. The schedule design problem involves selecting an optimal set of flight legs to be included in the schedule.

After market selection and appropriate forecast of passenger demand, the **airline's share** in all selected markets should be assumed (Holloway 2008).

Appropriate **flight frequencies should be determined** for the selected markets and the estimated market share. Special attention should be paid to the relationship between the size of the aircraft and the frequency of flights, for a predefined quality of service, market share, price, and load factor. Many authors in their papers researched this relationship (Givoni and Reitveld 2009, 2010, Pitfield et al. 2010, Wei 2006, Wei and Hansen 2005, 2007). These authors independently concluded that airlines would rather increase the frequency of flights than the size of aircraft, to keep their position on the market and respond to the competition. It is important to note that a high frequency between two cities does not always mean an increase in market share. If the increase in frequency on a route is not accompanied by an increase in the demand, the airline may perform a large number of flights with a low load factor. The frequency between city pairs (*FREQ*) can be determined as the ratio of the expected number of passengers (*PAX*) and the effective capacity of the aircraft ($FREQ = PAX/(CAP * LF)$ (Bazargan 2010)), where the effective capacity is calculated as the product of the average aircraft capacity (*CAP*) and the average load factor (*LF*). The average load factor on a route plays an important role in determining the flight frequency on that route. Load factor is influenced by flight times, frequency, type of service, and of

Table 6.4. Load factors and expected revenues.

Average fare	Expected number of passengers	Load factor	Expected revenue
$240	100	0.67	$24,000
$220	115	0.77	$25,300
$200	128	0.85	$25,600
$180	140	0.93	$25,200
$160	150	1	$24,000

Source: Bazargan 2010

course, fare levels (Bazargan 2010). The relationship that exists between the price and the load factor implies that a large value of the load factor does not necessarily mean maximum revenue. An important condition is the cost-effectiveness of the flight, i.e., reaching a load factor that is at least equal to the break-even load factor. Table 6.4 gives an example of expected revenue for different average fares and load factors. It can be noticed that the maximum income is realized when the load factor equals 0.85, while the income from a full aircraft is somewhat lower.

Operational constraints related to aircraft that are candidates to perform flights on considered routes and with defined frequency should be considered in terms of their range and possibilities to carry a payload. Additionally, airports should be considered in terms of runway length, obstacles, as well as their possibility to accommodate passengers and baggage (terminal, check-in counters, capacity).

After determining the flight frequency, the appropriate **departure times** should be determined. Departure times should be determined for each flight considering the market needs, whether the airline is a member of an alliance, whether its flights are connected to the following flights, as well as the airport slot availability. In addition to all the above mentioned, one should keep in mind different time zones, especially in long-haul flights, which can affect the departure time determination. Flights must be completed at a time when destination airports are open for aircraft operations. If the airline operates through hubs, i.e., if its flights are connected, departure times should be coordinated with departing flights from the hub to attract as many passengers as possible.

Passengers are more sensitive to changes in departure time on short-haul routes in comparison to long-haul routes. The reliability of a flight schedule also depends on the defined departure times. Moreover, inadequately determined departure times can result in long delays and/or flight cancellations that occur frequently. Thus, for example, meteorological factors at a certain airport can affect the operating hours of the airport, and thus the aircraft operations to a greater or lesser extent.

If meteorological factors are not considered when determining departure times, some flights may always perform with a delay during a certain period of the year (e.g., winter season). Departure times defined in this way can affect the loyalty of passengers, especially business passengers, for whom the reliability of the flight schedule is very important. Flight schedules are very sensitive to changes in departure times. For this reason, the problem must be addressed in the right way, so that the designed flight schedule can be sold on the market and the airline can attract as many passengers as possible. For example, changes of two hours in a departure time may result in the reduction of the number of passengers even by 50 percent, while shifting a departure time by 20 minutes may cause the airline to lose 10 passengers per day (Wensveen 2015).

The next steps in flight schedule development refer to the **calculation of block time** for each leg and **turnaround time** (Holloway 2008). To ensure that the designed flight schedule will be sold, block time[6] and turnaround time[7] need to be calculated properly, due to the fact that these times have a direct influence on schedule reliability, which further influences passengers' loyalty (especially loyalty of business passengers).

Block time of the observed flight may vary. Sometimes this is caused by meteorological conditions on the route being flown, or the traffic situation at the origin and destination airports. Therefore, certain problems can occur when calculating block time (Holloway 2008). One of the problems is the duration of the flight. The shorter the flight, the shorter time for potential recovery from irregular situations. Delays on shorter flights represent a higher percentage of the total block time than on longer flights.

When it comes to the time the aircraft spends on the ground, it is evident that airlines tend to minimize it to maximize aircraft utilization. However, minimal turnaround time has a negative impact in case of disturbances, because there is no possibility of delay amortization, and the disturbance will cause knock-on delays which propagate through other flights. This delay propagation reduces the reliability of the flight schedule. The task of airline planners is to design a flight schedule that will be robust and reliable, whereby the time the aircraft spends on the ground will be determined in a way to enable a reliable schedule and appropriate utilization of the company's resources (Figure 6.12).

Taking into account the characteristics of hub-and-spoke and point-to-point networks, as well as business policies of low-cost and traditional/regional airlines, it can be seen that turnaround time in hub-and-spoke

---

6 Block time is the time between engine starts at the airport of origin and engine stops at the airport of destination.

7 Turnaround time is the time that passes from previous landing until next take-off, excluding taxi-in and taxi-out.

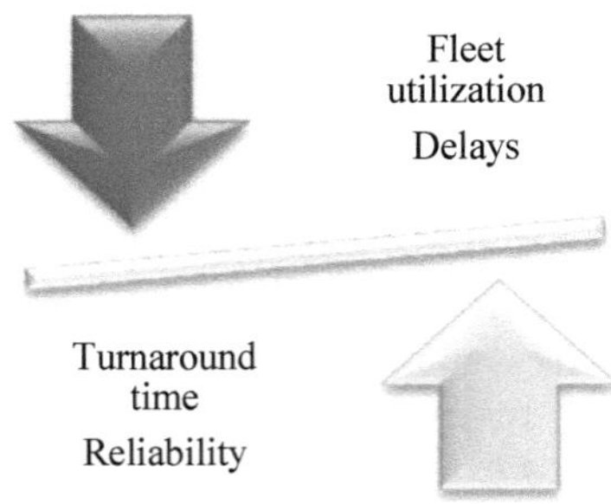

Figure 6.12. Turnaround time's influence on flight schedule.

networks is longer than in point-to-point networks. The LCC policy includes shorter turnaround times because they do not provide connecting flights. Moreover, the reduction of turnaround times for the airline means increasing block hours, i.e., increasing capacity expressed through ASM/ASK without changing the fleet (Holloway 2008).

Once block time and turnaround time have been calculated, the **fleet assignment** should be done (Holloway 2008). The fleet assignment problem involves assigning aircraft types (or fleets) to flight legs to maximize revenues and minimize operating costs simultaneously. The assignment depends on aircraft characteristics, route, and network, passenger demand, and airports where the airline flies to. In hub-and-spoke networks, larger-capacity aircraft are usually deployed on routes between hubs, while smaller-capacity aircraft perform traffic on shorter lines to hubs (Bazargan 2004). The fleet to be allocated to the route network can be homogeneous, which is the case with low-cost airlines that have only one type of aircraft in the fleet, and heterogeneous, which is the case of full service airlines that have different types of aircraft in the fleet. Each type of aircraft has certain characteristics (capacity, mass, maintenance, fuel, crew) and corresponding costs. A homogeneous fleet for an airline means advantages in terms of maintenance costs, buying of aircraft spare parts, labour, etc., but such a fleet is not suitable for operating on hub-and-spoke networks due to the impossibility to match supply and demand better on routes having different characteristics (e.g., length, demand). For such networks, a heterogeneous fleet is a better choice.

The next step in airline scheduling planning is to calculate the expected number of block hours of each aircraft type, considering all constraints related to **maintenance** (Holloway 2008). The primary goal of maintenance is safe traffic, i.e., providing a safe aircraft for each flight. It would be very easy to meet maintenance constraints if the airline had an unlimited number of aircraft, people, and equipment at its disposal at each airport (Wensveen 2015). As this is not the case, in reality, the problem becomes more complex. The airline strives to ensure that its aircraft spend minimal time on the ground, that the freely available time is used to the maximum for maintenance, that employees offer their maximal engagement with the appropriate workload, and that the airline's equipment is used to the maximum. In addition to all the above mentioned requirements, it should be ensured that the appropriate aircraft (the one to be maintained) is at the right time (maintenance schedule time) at the right place (airport where the required maintenance is performed) where all prerequisites for performing maintenance (equipment, manpower, maintenance plan) are provided. Therefore, the planning department needs to harmonize the proposed flight schedule with the maintenance department to provide a flight schedule that is feasible from the aspect of technical requirements and maintenance.

The aircraft routing problem (in the literature also known as aircraft rotation or tail assignment) is the problem that refers to assigning a set of flights to a set of aircraft to create a feasible flight schedule for an airline. This is a very complex problem that is being studied separately. In the academic literature, maximum income or minimum operating costs, the minimum number of aircraft, the highest aircraft utilisation before going to technical maintenance, etc., can be used as criteria functions. Flights covering, similar aircraft utilisation for all aircraft in the fleet, technical maintenance (airports where maintenance is performed), etc., can be used as constraints in the problem of aircraft routing.

The next step in flight schedule development concerns **crews**. More precisely, two main tasks related to crews can be observed, and they are crew pairing (determine the crew rotations) and crew rostering (assign crew members to corresponding duties (Holloway 2008)). While solving the crew rotation problem, the minimum number of crews required to perform the planned traffic is determined, while when assigning crews to duties, duties are assigned to each individual crew member (by name and surname) usually for a one-month period. Problems related to crews (crew pairing and crew rostering) are very complex, combinatorial problems, which are, similarly to fleet assignment and aircraft routing, solved separately and which have been widely studied in the academic literature. The goal of the airline in solving crew problems can be to minimize the number of required crews, the number of crew members, crew costs while meeting all regulatory requirements related to the allowed working hours of crew members.

An airline needs to permanently monitor changes in the market as well as its competition in highly competitive markets to predict **competitors 'strategies** and reactions (Holloway 2008), and to be able to react in the right way. Sometimes it is necessary to develop different plans (or if-then scenarios) to allow the possibility to choose a plan that fits the considered situation best.

Taking into account all previously described steps, it can be concluded that the process of flight schedule development is an extremely complex process. The flight schedule designing on a certain route network, by itself, is a procedure for determining the departure time for all destinations in the network, so that the given objective function reaches the optimal value, while all constraints are satisfied. The designed and adopted flight schedule represents an obligation for the airline. The operational success of the airline and its quality of service depends on the flight schedule, which is designed to fulfil passenger demand, utilize available resources, and satisfy different operational requirements. Passengers' and airlines' interests, which are usually opposing, should be harmonised and satisfied.

Airlines want to achieve the highest possible load factor, as well as aircraft utilisation, to have the lowest possible costs and the highest possible income (Figure 6.13). On the other hand, passengers want as many direct flights as possible, high values of flight frequencies, good connections, appropriate departure times, short waiting times, and a high probability of finding a free seat on the desired flight (Figure 6.13). Technical (operational) and regulatory constraints, which must be met, comprise aircraft technical maintenance planning, crew planning, airport opening hours, time zones, etc. The goal of every flight schedule is to provide the best possible offer to passengers, with appropriate reliability and punctuality. In addition to the feasibility in terms of operational constraints, the flight schedule must also be economically viable.

Therefore, it can be concluded and emphasized once again that this is one of the most complex and challenging tasks for an airline, and it cannot be completely mathematically formulated. To solve the problem successfully, the problem is divided into smaller sub-problems (frequencies, flight schedule, aircraft rotations, crews, maintenance) which are solved sequentially or simultaneously. The optimal solution

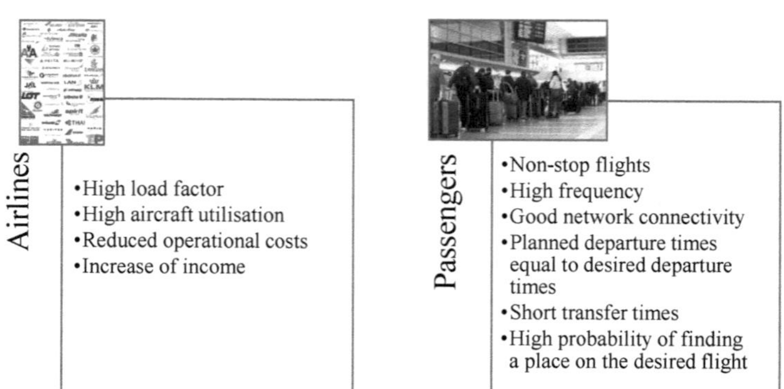

Figure 6.13. Airline's interests vs. passengers' interests.

to individual problems can, to a greater or lesser extent, corresponds to the optimal solution of the whole system. Therefore, the problem of flight schedule planning can be considered to be an iterative procedure that involves permanent coordination between the services involved in flight schedule planning.

In the process of airline flight planning, it is possible to distinguish between activities that are performed at the strategic level and those that are performed at the tactical level. Strategic development responds to the major changes in both business and operational environments and relates to the long-term planning of future route networks (Bazargan 2010). These activities are carried out over a period of several years in advance. Activities involved in long-term planning relate to fleet diversity, manpower planning, hub planning, and hub development, as well as appropriate facilities and equipment at airports (Figure 6.14). Market evaluations are performed for a period of 3–1 years in advance and comprise determining the frequency of flights in all markets, the possibility of expanding/reducing the market, defining the pricing policy, predicting competitors' behaviour, and decisions related to joining alliances and code-share agreements (Figure 6.14). In the next phase, the flight schedule is optimized (12–3 months in advance) by designing an initial flight schedule for the available fleet, then assigning the aircraft to flights and evaluating facilities and manpower (Figure 6.14). Tactical development refers to a short period of time (4–1 months) and it changes in this period. Operational flight schedule changes refer to crew changes, departure and landing times, maintenance, and they occur on a daily basis.

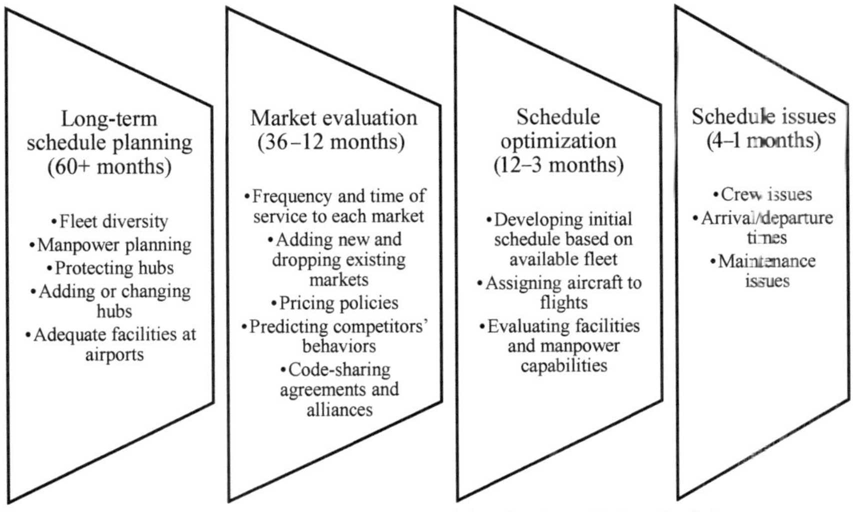

Figure 6.14. Activities by phases of developing a flight schedule.

A flight schedule is designed for a certain period of time, for example, for a season lasting six months. This flight schedule is usually called a seasonal flight schedule and it can be:

- winter schedule (October–March), and
- summer schedule (April–September).

The seasonal flight schedule of an airline is published twice a year and provides passengers with information on departure times during each day of the week at all destinations. The airline is obliged to adhere to the published flight schedule, which means that the planned flights must be performed regardless of the number of passengers on board.

During the season, various changes occur so the planned flight schedule is updated monthly and weekly. The last changes are entered 24 hours before the start of the flight schedule for the observed day, and such a flight schedule is delivered to the dispatcher in the airline operations centre.

The daily operational schedule is the final version of the flight schedule for the considered day, which takes into account all the changes that were undertaken during the period from its publication until just before its realization. The daily schedule, handed to the dispatchers at the airline operations control centre in charge of observing its realization, is a set of aircraft and crew (flight and cabin) routings for a one-day time period. For each planned flight within that time interval, there are defined departure and arrival times, departure and arrival airports, flight numbers, type, and registration of aircraft, which, according to the schedule, is to execute the given flight.

6.8.2 Operations and disruption management

As the flight schedule is realised, disturbances that cannot be predicted by the airline in advance occur. Dispatchers in the airline operations centre (AOC) are on a daily basis faced with disturbances caused by various factors, such as meteorological conditions, aircraft failure, crew absence or delay, errors in the estimation of block, or turnaround times at certain airports, airport congestion, etc. When disturbances make it impossible to realize the planned flight schedule, the dispatcher at the AOC defines a new daily operational flight schedule based on available resources. Consequences of flight schedule disturbances are flight delays and/or flight cancellations that lead to additional costs, loss of passengers' loyalty, damage to the airline's reputation, etc. (Figure 6.15).

Flight schedule disturbances can be classified as internal, which means that the airline can influence their solution. Such disturbances are a consequence of crew delays or crew absence (when the airline can hire a reserve crew), a technical failure of the aircraft (when the airline can

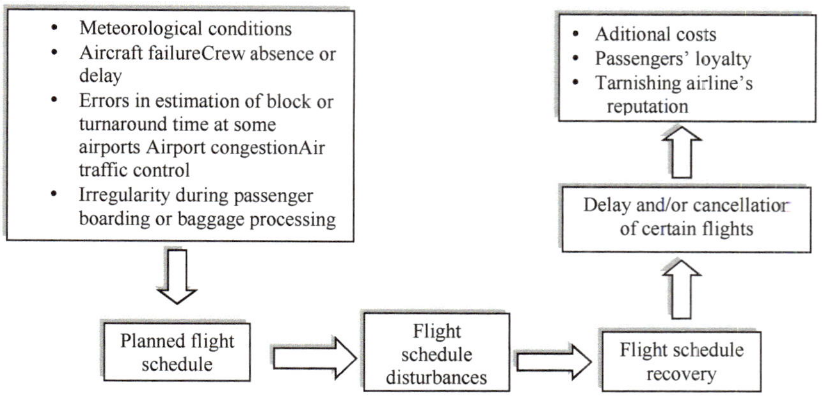

Figure 6.15. Airline schedule disturbance scheme.

repair the aircraft or hire a spare aircraft), etc. Disturbances can also be external, i.e. the airline cannot influence their solution. These are usually disturbances that occur as a result of bad weather conditions (in this case, the airline can only wait for weather conditions to change), congestion at the airport or in the air (the airline must follow the flight control instructions), etc.

Main actions that can be taken by dispatchers while solving the schedule disturbance problem are as follows:

- flight delaying, which directly affects passengers on that particular flight, and indirectly the passengers on the following flight in the rotation of the considered aircraft;
- aircraft swapping, which means that the aircraft can serve the flight which is not in its original rotation, if the capacity matches the number of passengers on the given flight;
- flight cancellation, which is an extreme option both for passengers and the airline, and it may cause serious disturbances;
- ferry flight, which means flight without passengers, and
- introducing a spare aircraft which is possible only if the airline has spare resources.

During the recovery of airline schedules, the dispatcher at the AOC considers the airline's policy and defines a new flight schedule that will optimize an objective function and satisfy the corresponding constraints. Depending on the airline's policy, when designing the new flight schedule, the dispatcher delays or cancels some flights and/or reassigns flights to available aircraft to reduce the negative effects of the disturbance. To minimize the negative effects of disturbances, the dispatcher, as a decision maker, is expected to create a solution and take appropriate action to implement that solution within a very short time period, i.e., in real time. The complexity of the disturbance problem is rising when the dispatcher has to deal with more flights and rotations. This is why, as a matter of major importance, the need for computational methods and techniques arises. Implementation of these methods and techniques in the decision-making process should facilitate the creation of feasible and economical solutions within an acceptable time limit. The process of airline rescheduling and recovering consists of interaction between human experts (dispatchers in the airline control centre) and the decision support system (DSS). The main goal of a DSS as a computer-based information system is to simplify the decision-making process through graphical user interfaces, to assess current schedules or elements, to check for feasibility of solutions regarding restrictions, and to automatically solve the problem in case of a disturbance. DSSs help dispatchers in their work by accelerating the decision-making process and increasing the quality of decisions made.

6.9 Airline economics

6.9.1 Airline cost structure

Airline costs can be divided into operating and non-operating costs, where the former are incurred due to the performance of air transport operations, while the latter are incurred due to the performance of non-air transport operations. Non-operating costs may vary with different airlines as a result of different activities and financial structures within the company. Operating costs arise from the production and sale of air transport services and can be categorized as direct or indirect costs.

Direct operating costs (DOC) are costs associated with and dependent on the type of aircraft being operated. Accordingly, DOC change if the structure of the fleet changes. The following costs belong to DOC: flight operation costs, maintenance, and overhaul costs, and all aircraft depreciation costs (Wensveen 2015).

Flight operation costs comprise the following costs:

- Flight crew salaries and other costs (travel costs, allowances, pensions, and insurance)—they represent the most significant costs of the airline due to higher salaries for crews in comparison with other workers. The pilot's salary depends on the type of aircraft being flown, and the general rule is that the larger the aircraft the higher the salary is paid. Costs associated with flight crews can be calculated on a route-by-route basis or they can be expressed as an hourly cost per aircraft type (Doganis 2010).
- Fuel and oil costs—these represent the second major cost element of flight operations. Fuel cost varies by aircraft type, depending on the

number, size, type, and age of engines. Moreover, during a particular flight, fuel consumption may vary in relation to the sector lengths, the aircraft mass, wind speed, cruise altitude, etc. (Doganis 2010). Fuel costs are also very sensitive to changes in the fuel market price and this is why many airlines adopted strategies to hedge fuel prices. Airline's fuel hedging strategy attempts to manage fuel price risk through concluding a contract with aviation fuel market vendors, committing to purchase fuel and oil in the future at a predetermined earlier price. Fuel and oil costs include all relevant taxes and duties, such as taxes on fuel and oil charged by governmental units, and fuel throughput charges charged by some airports on the volume of fuel uplifted (Wensveen 2015).

- Airport and air navigation services costs—airport costs include charges in respect of landing, parking, security, turnaround, passengers, cargo, lighting, and noise charges. Air navigation service charges include en-route and terminal navigation charges. Airlines have to pay airport operators for the use of the runway and terminal facilities and this cost is calculated based on aircraft MTOM and the number of passengers boarded at that airport. Air navigation charges the airline has to pay to the air traffic control service provider to cover the cost of navigation aids that its aircraft use during the flight in the airspace under the competence of that air traffic control. This charge is calculated based on the aircraft's maximum take-off mass and distance flown.
- Aircraft insurance costs—these include the aircraft hull and liability insurance expenses. The hull premium is generally calculated as a percentage of the full purchase price of the flight equipment and may range from 1 to 2 percent, or lower. This depends on the airline, the number of aircraft the airline has insured, and the geographic areas in which its aircraft operate (Wensveen 2015). Liability premiums are generally based on the estimated number of revenue passenger miles flown. Additional coverages, such as war risk coverage or coverage against terrorist action, may be purchased for an additional premium (up to 2 percent) (Doganis 2010).
- Other flight-operations costs—these include costs not covered by the above mentioned categories, such as costs of flight crew training, costs of route development, rental or lease charges for hiring or leasing of aircraft from other airlines or leasing companies (Doganis 2010).

Maintenance and overhaul costs include a wide range of costs related to different aspects of maintenance and overhaul of the airline fleet. These costs depend on the fleet size, fleet age, network structure, aircraft utilisation, flight operational practice, and maintenance programme design. Maintenance input costs cover labour costs, material, and overhead costs. They cover routine maintenance and maintenance checks carried out between flights or overnight, as well as extensive periodic overhauls and major checks (Doganis 2010).

Aircraft depreciation and amortization costs relate to the depreciation of flight equipment on a systematic basis over its useful life to its residual value. Generally, each aircraft component should be categorised and depreciated separately using its specific useful life and residual value. Airlines' practice is to adopt the straight-line depreciation method over a given number of years, with a residual value of 0 to 15 percent (Doganis 2010). The depreciation period is determined by the aircraft type, and for wide-body jets, it ranges from 14 to 16 years, while for smaller short-haul aircraft it is generally 8 to 10 years (Wensveen 2015).

Indirect operating costs (IOC) do not depend on the type of aircraft, but largely on the number of passengers. These costs cover (Wensveen 2015):

- Station and ground costs—these represent the costs associated with all those services, incurred in providing the airline's services at an airport, such as salaries and expenses of the airline staff located at the airport and engaged in handling and servicing of aircraft, passengers, or freight; costs of ground handling equipment, of ground transport, of buildings and offices and associated facilities, and of communication equipment.
- Costs of passenger services—these are all the costs arising from passenger services, i.e., the payroll, allowances, and other expenses related directly to aircraft cabin staff and other passenger service personnel (e.g., hotel and other costs associated with overnight stops, as well as training costs of cabin staff).
- Ticketing, sales, and promotion costs—these costs include all costs associated with reservations, sales, and promotional activities, as well as all office and accommodation costs arising from these activities. These costs also include commission or fees paid to travel agencies and to global distribution systems for ticket sales, as well as to credit card companies for sales paid by card.
- General and administrative costs—these are cost elements that are general to the airline or which cannot be allocated to a particular activity or any above mentioned category.

Non-operating costs are not directly related to flight operations and generally can be divided into these categories: interests paid on loans, loss on asset disposal, costs arising from the airline's affiliated companies, costs arising from foreign exchange transactions, etc.

This is a common approach to airline cost classification, however, airlines tend to have different organizational structures, and thus they have different cost structures. This cost structure will even differ within the same type of airlines (they will have different cost allocations into DOC and IOC). Generally speaking, the percentage of DOCs in operating costs for charter airlines is up to 80, because they have minimal costs of ticketing, sales, and promotions. For scheduled airlines, the share of DOCs ranges from 45 to 60 percent of the operating costs, and this largely depends on the fuel market price, the average leg length, and labour costs.

Airline costs can be divided based on their relation to output into fixed, variable, total, average and marginal. Fixed costs (FC) of the airline do not vary with changes in the output level within the capacity range of the existing fleet structure. However, FC increase due to the expansion of the fleet, expansion of business premises, etc. Variable costs (VC) of the airline do vary with changes in the output level, and they shift in the same direction, i.e., VC increase with the increase in the output level, and decrease with the decrease in the output level. VC include fuel costs, air navigation and airport charges, regular maintenance costs, etc.

For fixed and variable costs, the time horizon in which they are observed plays an important role. In the long run, all fixed costs become variable, because the airline has an opportunity to change them by purchasing a new aircraft, developing new routes, etc.

When fixed and variable costs are added together, the result is the total cost (TC). When the airline has no service (no flight operations), then TC and FC are equal because there are no VC, i.e., there is no traffic to generate them. If the traffic is higher than zero, but still within the existing capacity, any increase in traffic will also increase the TC by the value of the variable costs incurred by the given traffic. The FC remain unchanged. Once the existing capacity is fully utilized, each subsequent increase in traffic levels generates certain fixed costs that will significantly increase the TC. If the new capacity is not fully utilized, the additional FC per unit of output will be higher. As the utilization of new capacity increases, the VC will increase, while the FC will decrease, i.e., they will be distributed over a larger number of seats. This process is repeated every time the capacity is used to the maximum and it is necessary to add new capacity.

In addition to total operating costs (TOC), there are two important cost categories: average and marginal costs. Average costs (AC) or unit costs are the total costs at a given capacity level divided by the total output. They are expressed as costs per ASK/ATK. Smaller aircraft generally have higher costs per seat compared to larger aircraft. This drop in costs per seat with the increase in the size of the aircraft is mostly due to the improvement of technology and structure of the aircraft itself (Holoway 2003).

Marginal costs (MC) represent the increase in TC that occurs due to the production of one additional unit of output. While MC are below AC, each additional unit of output will lead to a reduction in AC. When

MC exceed AC, each additional unit of output will lead to an increase in AC. Of course, this will only matter depending on what happens to the marginal revenue that refers to the increase in revenue that results from the sale of one additional unit of output. Profit is maximized at the point where marginal cost and marginal revenue are equal (Figure 6.16).

In the period after the deregulation of air transport market, airlines were under pressure to reduce their operating costs to be competitive and survive in the market. They had two options: cost-cutting strategies and becoming more cost-efficient. Cost-cutting strategies may include some or all of the measures related to laying off of excess employees, reducing the fleet size, restructuring the network, reducing the frequency, etc. Becoming more cost-efficient means that the airline needs to maximize the output for a given level of input (number of employees, number of aircraft, etc.). However, achieving efficiency in terms of operating costs does not mean that the airline needs to have the lowest operating costs, but it means that the incurred level of costs reflects the level of quality offered to passengers. Additional costs may be justified if they involve upgrading of a service that will attract more passengers who are willing to pay a higher price for that service.

The result of fierce competition and the deregulated market is that today almost all airlines have significantly lower costs compared to the period before the deregulation. Strong competition from LCCs has forced traditional airlines to change their operations or to withdraw from the market. Strategically, traditional airlines have reacted in different ways, but they all had something in common, and that is cost reduction. In addition to increased competition, the reduction in airline costs was also affected by one of the biggest economic crises in 2008 (Belobaba et al. 2009).

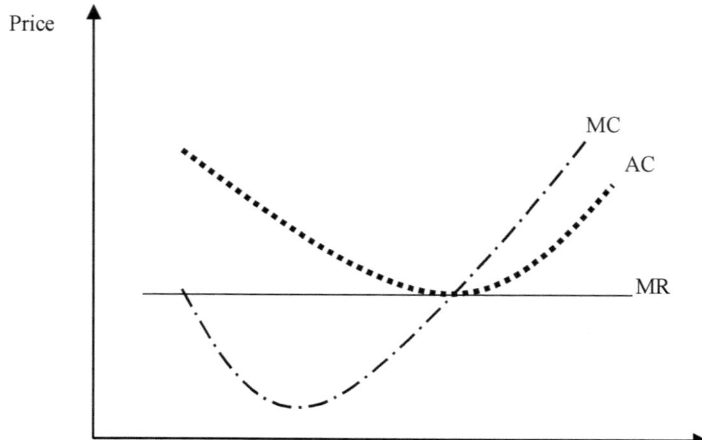

Figure 6.16. Average cost (AC), marginal cost (MC) and marginal revenue (MR).

Cost economies. Airline costs are affected by economies of scale, economies of density, economies of scope and average leg length. These factors can play a major role in the cost structure of the airline.

Economies of scale exist when unit costs decrease due to an increase in the output volume, while resource prices stay unchanged. In the airline industry, economies of scale exist when the number of produced capacity (ASK) increases, while the unit cost per ASK declines. ASK can increase by increasing the network size, frequency on the routes, etc. Economies of scale can also have a negative impact. Namely, the increase in capacity, which is not accompanied by an adequate increase in demand, will result in a decrease in the load factor and a decrease in revenue. Also, adding capacity implies certain fixed costs, so that the additionally produced capacity not only has to be sold, but must be sold at a price that will cover those additional costs.

Economies of density exist when unit costs decrease as a result of an increase in traffic volume on a particular route already served by the airline, while the network structure remains unchanged. This can be achieved by adding a flight or using a larger capacity aircraft. Economies of density are closely related to aircraft size, due to the fact that the relationship between increase in aircraft size and increase in flight operating costs is not linear. If the airline replaces the smaller aircraft with a larger one on a certain route, this will result in a reduction in unit costs per seat. There are several reasons why this relationship is not linear: a larger aircraft does not necessarily have higher fuel consumption per seat-kilometre compared to a smaller aircraft, a larger aircraft does not require more flight crew members, nor proportionally higher maintenance. Therefore, one of the ways to exploit the economies of density would be to design such a network structure that would channel traffic from certain routes into a hub to increase the number of passengers on each route. In this way, a sufficiently large demand would be created to justify the use of a larger aircraft with lower unit costs.

Economies of scope exist when unit costs decrease if one airline produces two different products in comparison to the case when they would be produced by two airlines individually. Applied to the airline network, economies of scope exist when passengers who travel from/to different destinations fly at least one part of the route on the same aircraft. In other words, when seats in a single aircraft, which carries passengers from point A to point B, are sold to passengers who started their journey before point A, who continue their journey after point B, as well as to passengers who travel from A to B. Economies of scope can also occur when the airline introduces a non-stop flight between points that it normally serves indirectly over a hub. Economies of scope exist when it is cheaper to offer different types of services on one flight, instead of offering each of the services on different flights. The most common example of this type are the first, business and economy class offerings in the same passenger cabin on the same flight, where all services share the same fixed costs and there is an additional benefit of economies of density because a larger aircraft is used compared to the size used for only one service offered on the flight. Moreover, economies of scope exist when passengers and cargo are transported on the same flight.

The average leg length also affects the costs of the airline and its influence is in reducing the unit operating costs as the length of the leg increases, all other things being equal. There are several reasons why unit operating costs decrease with increasing leg length, viz. (Holloway 2008):

(a) The time an aircraft spends on the ground for turnaround is very short and it decreases compared to the time the aircraft spends in the air, as the average leg length increases. This leads to an increase in crew and aircraft utilization;
(b) There is an increase in average cruise speed compared to the speed on shorter legs, which increases the productivity of the aircraft per hour and thus increases the unit capacity (ASK) over which fixed operating costs are allocated;
(c) Fuel consumption per kilometre decreases the longer the aircraft stays in the optimal cruising speed and at the optimal altitude;
(d) The costs of regular maintenance are higher for short-haul aircraft, because they accumulate working hours faster, after which regular maintenance is mandatory.

Although long-haul flights have an advantage in significantly lower costs, this does not mean that airlines operating on shorter routes are necessary at a disadvantage. They compensate their higher unit costs with higher ticket prices and generally have higher revenues per kilometre flown compared to airlines that operate long-haul flights. Also, the lack of higher unit costs is offset by lower in-flight service costs because shorter flights require less service during the flight.

6.9.2 Airline pricing

The ticket price has always been one of the key choice factors influencing passengers' decisions which transport mode to use. As the air travel demand is defined for an origin-destination (O-D) market, the airline ticket price is also defined for the O-D market, not for an airline flight leg.

Pricing in airline industry refers to the process of determining the price levels (for passenger and freight), combined with various service amenities and restrictions, for a set of products in an origin-destination market (Belobaba et al. 2009). Airline prices for each O-D market depend primarily on the volume and characteristics of the demand, the nature of airline supply (route, aircraft size and frequency) and the number and type of competitors on that market. The pricing process also involves determination of the rules and conditions applicable to each price to be offered. Each price paired with conditions constitutes the so-called, 'tariff' or 'fare'. Fares will be explained later in this chapter.

The air ticket price consists of two parts. The first part represents the airline revenue and it is charged for the transport service. In air transport, the ticket price should cover fuel costs (which are generally directly proportional to the distance the aircraft flew), costs related to employees' salaries, routine inspection of aircraft, in-flight services provided to passengers, catering costs, depreciation, marketing and administrative costs. The second part of the ticket price covers the airport charges for services provided to passengers by the airport operator, which the airline is legally obliged to charge to its customers and pass on to the airport. In general, airlines pay the following charges after arriving at the airport: for aircraft landing, for the use of lighting systems, for parking, parking in the hangar and long-term parking, for passenger services in domestic or international traffic, or for passenger services at different terminals at the same airport and aircraft turnaround. In addition, at some airports, airlines have to pay charges for noise and for security. What is directly included in the ticket price, which must be clearly pointed out, is the part of the airport charges that refers to passenger services and the part that refers to the security service. Therefore, the total price of an air ticket is the sum of the price, airport charges for passenger services, security charges, and other charges.

To survive in the new market conditions (market liberalization and airport privatization), many airports around the world have offered certain discounts on airport fees charged to airlines. Over the last decade, several airports in the world have developed various types of measures aimed at encouraging the growth of traffic. Today, there is a whole range of various incentive programs for airlines to generate additional traffic. Possibilities of incentive measures for airlines are practically limited by the conceptual design of the entity that creates them and the regulatory environment that must always be met.

In mid-2019, due to the introduction of a special security charge, passenger charges at Belgrade airport were increased from € 16.5 to € 21 at Terminal 1, while at Terminal 2 the charges were increased from € 12 to € 16.5. The charge for transfer passengers at Belgrade airport is € 4. In Europe, airport charges range between € 5 and € 35 per passenger (data from 2013). It is clear that small (secondary) airports that want to attract airlines have significantly lower charges. For example, Memmingen Airport in Germany charged the airport fee of € 5 in 2013, while in 2019 it was slightly more than € 7. If a passenger flies from Belgrade to Zurich in one direction, the ticket price will include the departure airport charges of Belgrade airport. If a passenger flies on the Belgrade–Zurich route with a return ticket, then the ticket price includes the departure airport charges for Belgrade airport and the departure airport charges for Zurich airport (in return). If the passenger flies on the route Belgrade–Zurich–Washington in one direction, he/she needs to pay the outgoing Belgrade airport charges and the transfer airport charges at Zurich airport. If the passenger flies on the route Belgrade–Zurich–Washington and back, then he/she needs to pay the outgoing Belgrade airport charges, outgoing airport charges in Washington (on the way back) and twice the transfer charges at the airport in Zurich.

For example, for a return ticket Belgrade–Athens (Air Serbia), with departure on November 29, 2019, and return on December 4, 2019, the total price of the air ticket was € 172, of which the airline ticket price was € 116. The difference of € 56 refers to different airport charges, where about 40 percent belongs to Belgrade airport, and 60 percent to the airport in Athens. In this case, one third of the ticket price was charged for airport charges.

The ratio between the airline's price and the charges within the air ticket price has changed in favour of charges. According to IATA,[8] airport charges doubled in 2016, compared to 2006. In absolute terms, if the airport charges in 2006 were € 16 per passenger, ten years later they were € 33 per passenger. If other charges are added to airport charges, the difference is not negligible (in 2006 it was € 40, and in 2016 it was € 82). Regarding the fare in the same period, they were lower on average by 27 percent.

When establishing the prices, an airline may follow one of three different theoretical bases for determining the prices to charge in an O-D market (Belobaba et al. 2009):

- Cost-based pricing—generally refers to a principle when the airline sets its prices in all O-D markets based on average operating cost per flight or per available seat kilometre (ASK). This was the pricing principle used during the period of air transport regulation.
- Demand-based pricing (price discrimination)—refers to a principle that a passenger is charged based on his/her willingness to pay, in each O-D market. This way of pricing results in different prices for different O-D markets, even for different demand segments within the same market. The price differences are not related to cost differences in providing services to the different demand segments, only to the differences in price sensitivity, demand elasticity and willingness

[8] https://blueswandaily.com/taxes-charges-and-fees-are-pushing-up-airline-fares-to-the-detriment-of-passengers-argues-iata/, 25.12.2019.

to pay. This is the reason why this principle is considered price discrimination since the airline is charging different prices to different customers for the same product.

- Service-based pricing (price differentiation)—refers to a principle that the airline uses to charge the product based on differences in the quality of services (e.g., first class vs. economy class). Service-based pricing has a differential cost basis for the airline, because higher-quality services generally cost the airline more to produce.

In practice, most airline pricing strategies reflect a mix of these theoretical principles.

Depending on the market conditions and on the airline's policy, more than 30 different fares can be defined on one flight. Fares charged by the airline reflect the class of flight, travel time (peak hour or not), restrictions related to the reservation, and in some markets may reflect the choice of route (non-stop flight or flight with a connection at the hub). Many airlines grouped fares in classes where differences between fares from the same class should be minimal, while differences between fares from different classes should be maximal. Fare structure differs from airline to airline, as well as from market to market. However, some general concept can be observed (Figure 6.17).

Published fares are available to anyone for purchase and they are usually called public fares at Airline Tariff Publishing Company, ATPCO, and SITA's AIRFARE system. These fares can be divided into unrestricted (flexible or full) fares and discount fares. Full fares are fares without any restriction—they are flexible. These tickets can be refunded or changed as necessary, and can be purchased on the day of travel for all classes on the flight (reservation classes: P—long-haul flight first class, F—short-haul/

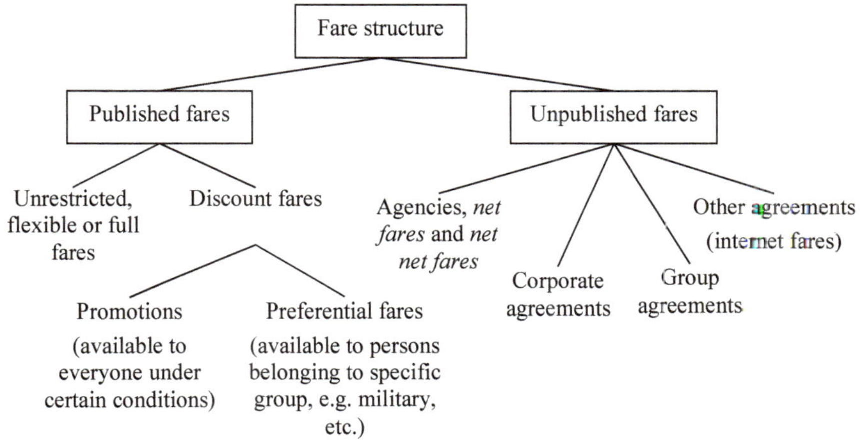

Figure 6.17. Fare structure.

medium-haul flight first class, J—long-haul flight business class, C—short-haul/medium-haul flight business class, Y—economy class). Therefore, these are the most expensive types of airfares, and they represent the regular price of the fare.

Discount fares are cheaper than full fares and they are offered in each class and in almost all markets. They are offered under certain conditions and restrictions, and they are usually available for a limited time. Discount fares can be promotional (occasional discounts), available to anyone who is ready to meet the defined requirements (when the fares can be purchased, minimum and maximum stay) or preferential fares (permanent discounts) that are available to people who meet the defined conditions specified by an attribute or employment status. Categories of passengers entitled to preferential fares are:

- INFANT—usually classified as up to 24 months old (2 years) and without a seat;
- CHILD—a person who has reached their second birthday, but not his/her twelfth birthday as of the date of commencement of travel from the journey origin;
- YOUTH—a person who has reached their twelfth birthday but not his/her twenty-fourth birthday as of the date of commencement of travel from the journey origin;
- SENIOR—a person older than specified (older than 60, or older than 65, depending on the airline);
- Airline's employees;
- Different professional groups—students younger than 27 and holding the International Student Identity Card (ISIC), professors with the International Teacher Identity Card (ITIC), sailors, journalists, etc.

Unpublished fares are not available on general sales channels (online through flight booking websites or on the airline's own website, neither can be booked by airlines). These fares are accessible by travel agents through the Global Distribution System (GDS), or can be classified as *Net* or *Net net* fares. Net fares refer to the total amount the agency pays for a block of seats reduced by the standard commission and any other price adjustment that is typical of a given market. A net net fare refers to a net fare reduced by any incentive the agency receives from the airline to ensure loyalty. All other unpublished fares refer to different types of agreements between airlines and companies, governments or other organizations, where the fares are negotiated between the stakeholders.

To be profitable and to gain as large profit as possible, airlines need to limit the use of discounted fares, and this can be done by effectively employing 'fences' or barriers. Namely, airlines want to enable business passengers to pay the most expensive fares rather than the discounted ones. Moreover, airlines can achieve this goal using the following fences (Vasigh et al. 2018):

- Advance purchase restrictions—these restrictions are related to the fact that price-sensitive passengers book air tickets more in advance, while time-sensitive passengers book their ticket later; therefore, earlier booking is related to lower fares (Table 6.5);
- Saturday Night Stay—business travellers usually book Monday morning flights and Friday afternoon flights; therefore, this fence was introduced as an aid to market segmentation (business and non-business travellers); as trends in business passengers' behaviour change, many low-cost airlines have abolished this barrier (Table 6.5);
- Frequent Flyer mileage—frequent flyer miles can be calculated in different ways, and depend on the fare class; sometimes passengers will be willing to pay a higher fare class to earn more miles (Table 6.5);
- Ticket refundability—this barrier is another aid in market segmentation; higher fare classes are fully refunded, which gives time-sensitive passengers more flexibility; some lower fares are refundable only under certain conditions, and some are not refundable at all (Table 6.5);
- Change fees—higher fare classes allow changes of the flight without additional fees (a business passenger can make last-minute changes if his/her schedule is changed, free of charge); in lower fare classes, a change of flight can be made for a certain fee, while in some fares change is not possible at all (Table 6.5);
- Schedule-driven fences—business travellers usually travel early in the morning and return late in the afternoon; for this reason, airlines should limit the number of low fares on these flights.

Table 6.5. Examples of fences.

Fare class	Advance purchase restrictions	Saturday night stay?	Frequent-flyer mileage	Refundable?	Change fee
Y	No	No	150%	Yes	Free
M	7-days	No	100%	Yes	$25
Q	14-days	Yes	100%	No	$150
T	21-days	Yes	50%	No	No changes

Source: Vasigh et al. 2018

6.9.3 Revenue management

Revenue management is based on the concept to sell tickets to the right passengers at the right time and for the right price. A seat in an aircraft that is left unsold is impossible to store for future sale, and thus the seat is considered to be perishable. All seats that remain unsold before the departure of the aircraft cannot be returned and sold again after the departure. Therefore, unsold seats increase costs and reduce the revenue of a particular airline. The airline's revenue also depends on the price of seats. Although charging all seats on the aircraft with the lowest price may result in a filled passenger cabin, this does not necessarily mean the highest revenue for the airline. It can happen that the highest revenue is realized with a lower load factor and higher prices. The revenue management system is a very important strategic tool that allows an airline manager to handle the supply of aircraft seats and passenger demand to maximize the total airline revenue.

The basic idea behind airline revenue management is that it may be better not to sell a seat today at a low price if it can be sold tomorrow at a higher price, or allowing the seat to be sold today at a low price if it is otherwise likely to remain unsold (Vasigh et al. 2018). Revenue management has had great success in the airline industry due to the specifics of the business environment characterized by:

- High fixed cost for offering the product (aircraft capacity costs), while the marginal costs are generally low (e.g., fuel and other flight-related costs);
- Fixed offered capacity that is determined by the fleet (size and structure) and by the published schedule;
- Perishable products in terms of offered seats that are impossible to store and, thus, any seat or cargo space that is left unsold before the departure will be lost;
- Uncertain demand that varies in time of the day, day of week and/or season;
- Demand that can be segmented in many different ways, i.e., different passengers must be willing to pay different prices for the same service;
- Availability of relevant data to be able to forecast the demand by segment and by departure.

The capacity on board of each aircraft is divided into a number of booking classes within each travel class (business, economy, first, etc.) that differ in terms of service (rebooking options, advance booking periods, etc.) and price. The goal of the revenue management system is to optimize the number of seats at different booking classes that are sold before each departure to achieve the maximum revenue. Revenue management comprises a pricing strategy and seat inventory control. The result of the pricing strategy is the number and type of different prices (tariffs) that will be offered in the market for each flight. Tariffs on flights vary from very high to very low prices. Full-service airlines offer the lowest tariffs to be competitive in the market to low-cost airlines. In this regard, the important issue related to the seat inventory control is how many seats the airline should offer at low prices.

The purpose of seat inventory control is to determine the number of seats to be allocated in each of the defined booking classes to maximize the total revenue. Assuming that prices have already been determined (based on costs, passenger willingness to pay and competition), the revenue management problem has been largely reduced to the seat inventory control problem. In essence, it is a probabilistic problem, due to the uncertain number of passenger requests for a certain class on the flight, a certain route and prices.

Revenue management, as mentioned, is concerned with predicting the demand and managing price and seat availability, to maximize the revenue, so below there are some guidelines for creating an effective revenue management system (Vasigh et al. 2018):

- Establish a differential price structure;
- Determine a system of constraints for using lower-tariff seats to limit the use of discounted seats to passengers who might otherwise be willing to pay much higher tariffs (e.g., advance purchase requirements, Saturday night stay, ticket refundability, frequent flyer mileage, etc.);
- Determine a system of seat allocation which maximizes expected revenue in respect of stochastic demand;
- Acquire tools for forecasting the demand, no-shows, cancellations, overbooking, and inventory limit.

Overbooking has become a regular airline practice when airlines, in some cases, book more passengers on a flight than the number of seats available. This process is a part of the revenue management problem and allows the airline to manage its revenue better. Overbooking practice is the result of the behaviour of passengers who cancel their flight reservations at the last minute, or misconnect, so the airline is unable to re-sell their seats, or those passengers who travel on flexible tickets and may not cancel their reservations before the flight (called 'no-shows'). In both cases, the seats allocated to such passengers will remain empty and the airline will lose an opportunity to generate more revenue. A major issue in overbooking is to balance the anticipated revenue from selling extra seats versus the anticipated cost[9] of not having enough capacity to accommodate all the passengers (Vasigh et al. 2018).

There are two approaches in allocating seats by booking classes: nested and non-nested, Figure 6.18. Before explaining differences in these approaches it is necessary to define some key terms. Booking limit is the maximum number of seats that can be purchased for each booking class and it is applied to low-booking classes to protect the seat inventory in higher classes (Holloway 2008). Protection level is the number of seats that are left unsold so that they may be purchased by a higher booking class, (Vasigh et al. 2018).

In the non-nested approach, each booking class is assigned a certain number of seats (total number of tickets sold in a given booking class must not exceed that number), Figure 6.18 (Bazargan 2010). Under the non-nested approach, protection level and booking level are equal. The sum of the seats for each booking class is equal to the total capacity of the passenger cabin in the aircraft.

In the nested approach, each booking class is assigned the maximum number of seats that can be sold at that price (booking limit)

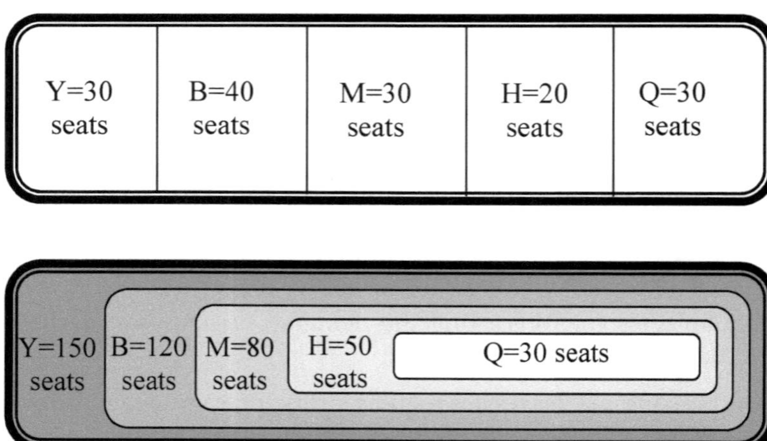

Figure 6.18. Non-nested and nested approach.

[9] Costs associated with overbooking include meal and hotel vouchers, flight coupons for future flights, departure delays, passengers being rolled over to other flights, staffing issues, and loss of goodwill. This happens when more passengers show up for a flight than there are seats available, so the airline will ask for volunteers to give up their seat. If no one is willing to give up their seat, some passengers will be denied boarding (Vasigh et al. 2018).

(Bazargan 2010). This means that for each booking class a limit is assigned for the maximum number of possible bookings, Figure 6.18. The limit of the number of possible bookings of a certain booking class is the sum of the number of seats assigned to that booking class and the number of all seats in the lower booking classes. In this case, protection level is calculated by simply finding the difference between each booking class. A request for a reservation in a certain booking class shall not be rejected in the event that the number of reservations reaches the number of seats allocated to that booking class, as long as there are free seats in any lower booking class.

In the example presented in Figure 6.18, the difference between the nested and non-nested approach to the seat inventory control can be explained in the case of booking class B. In the non-nested model, booking class B is allocated 40 seats and that is the maximum number of tickets that can be sold at that price. However, in the case of the nested model, booking class B is allocated 120 (40 + 30 + 20 + 30) seats.

The main disadvantage of the non-nested approach to the seat inventory control is that once all the tickets assigned to that booking class have been sold, any subsequent request to book a seat in that class will be rejected. In the case of the nested approach, this refusal will not happen, because the seat, if available, will be 'borrowed' from one of the lower booking classes, and sold at a higher booking class. The vast majority of airline reservations systems now have inventory structures based on nesting of booking classes.

Revenue management on one leg. Each flight in the airline schedule has a limited capacity, so the revenue management on one leg is to determine the number of seats that should be made available to each of the different booking classes available in the passenger cabin (e.g., the economy-class cabin) that maximizes the total revenue of each flight separately. To determine the number of seats in each booking class, the revenue management system must first forecast the future booking demand for each booking class on each departure. The next step is to estimate the revenue value associated with bookings in each booking class. The most commonly used approach to determine the number of seats in different booking classes is called *the expected marginal seat revenue* or EMSR approach, first introduced by Littlewood (1972), and later developed by Belobaba (1987). The EMR of potentially selling a seat in a booking class is the probability of being able to fill that seat multiplied by the average tariff of that class. They propose that as long as the expected marginal revenue from a seat for a higher booking class is larger than that of a lower booking class, then the seat should not be sold at a lower price.

Generally, revenue management for one flight needs to balance between attracting price-sensitive passengers who book early (through lower tariffs) and protecting the seats for business passengers who buy tickets closer to the date of departure and are willing to pay much more (through higher tariffs). The result of careful management of this process is that both load factors and revenue are maximised.

Revenue management on network. The seat-inventory control problem becomes more complicated when there are connecting and point-to-point passengers on the same flight, i.e., passengers on the same flight having different itineraries caused by hub-and-spoke systems. Generally, for an airline it will be better to accept requests for short-haul flight from passengers who are only flying out and back on the route (higher revenue per kilometre flown), than to accept requests from those who are using the short sector to fly to the hub, from where they will connect onto a long-haul flight (lower revenue per kilometre flown). On the other hand, if the airline rejects a connecting passenger because of a point-to-point passenger, this will improve its revenue result for that flight, whilst worsening the airline's overall financial results because of the loss of the so-called network revenue. This is why Airlines tend to apply origin-destination control in revenue management systems, to be able to manage their seat inventory by the revenue value of the passenger's origin-destination itinerary, not simply according to the booking class requested on a single flight leg (Belobaba et al. 2009). Revenue management for connecting passengers is different from that of point-to-point passengers in terms of more combinations of itinerary that need to be considered to decide how to fill each flight to maximize the network revenue. This implies a comprehensive analysis to be applied to evaluate the pricing on each origin-destination before deciding which markets to prioritise over others (Budd and Ison 2020).

Revenue management applied to the whole network needs to make sure that seats for connecting itineraries involving more than one flight leg are available in the same booking class across all flight legs requested by the high-revenue passenger. At the same time, the revenue management system is expected to be able to prevent lower-yield connecting passengers from displacing high-yield, short-haul or local passengers (Belobaba et al. 2009).

References

Basso, L., Clements, M. and Ross, W. 2009. Moral hazard and customer loyalty programs. American Economic Journal Microeconomics 1: 101–123.

Bazargan, M. 2010. Airline Operations and Scheduling. Ashgate Publishing Limited, England, Ashgate Publishing Company, USA, 2nd ed.

Belobaba, P. 1987. Airline Yield Management: An overview of seat inventory control. Transportation Science 21: 63–73.

Belobaba, P., Odoni, A.R. and Barnhart, C. 2009. The Global Airline Industry. Chichester, U.K: Wiley.
Bieger, T. and Wittmer, A. 2011. Airline Strategy: From Network Management to Business Models, In the book: Aviation Systems, 77–102.
Borenstein, S. 1991. The dominant-firm advantage in multiproduct industries: Evidence from the U.S. Airlines. Quarterly Journal of Economics 106: 1237–1266.
Budd, T. 2010. Environmental impacts and mitigations, Chapter 17. *In*: Graham, A., Papatheodorou, A. and Forsyth, P. Ashgate (eds.). Aviation and Tourism: Implications for Leisure Travel.
Budd, L. and Ison, S. 2020. Air Transport Management: An International Perspective (2nd ed.). Routledge.
Cento, A. 2009. The Airline Industry: Challenges in the 21st century, Physica-Verlag Heidelberg, A Springer Company.
Chin, A.T.H. 2002. Impact of frequent flyer programs on the demand for air travel. Journal of Air Transportation 7: 53–62.
Clark, P. 2007. Buying the Big Jets: Fleet Planning for Airlines, 2nd Edition, Ashgate Publishing Company, England, USA.
Doganis, R. 2001. The Airline Business in the Twenty-first Century, Taylor & Francis Group, Routledge, London and New York.
Doganis, R. 2010. Flying Off Course: Airline Economics and Marketing, Routledge, 4th ed.
Dožić, S. and Kalić, M. 2015. Three stage airline fleet planning model. Journal of Air Transport Management 46: 30–39.
Dožić, S. and Krnić, R. 2016. Overview of airline fleet ownership structure. Proceedings of XV International Symposium Symorg, 854–862.
Dožić, S. and Kalić, M. 2018. Aircraft type selection problem: application of different MCDM Methods, the advances in intelligent systems and computing series. Advanced Concepts, Methodologies and Technologies for Transportation and Logistics 572: 156–175. Springer.
Dožić, S. Lutovac, T. and Kalić, M. 2018. Fuzzy AHP approach to passenger aircraft type selection. Journal of Air Transport Management 68: 165–175.
Givoni, M. and Reitveld, P. 2009. Choice of aircraft size – explanations and implications. Journal of Air Transport Management 16(3): 159–167.
Givoni, M. and Reitveld, P. 2010. The environmental implications of airlines' choice of aircraft size. Transportation Research Part A 43(5): 500–510.
Graham, A. 2010. Trends and characteristics of leisure travel demand, Chapter 3. *In*: Graham, A., Papatheodorou, A. and Forsyth, P. Ashgate (eds.). Aviation and Tourism: Implications for Leisure Travel.
Hanlon, P. 2007. Global Airlines: Competition in a Transnational Industry, Butterworth-Heinemann; 3rd ed.
Holloway, S. 2008. Straight and Level: Practical Airline Economics. Ashgate Publishing Limited, England, Ashgate Publishing Company, USA, 3rd ed.
ICAO Glossary, Main Terms Used in Civil Aviation Statistics, https://www.icao.int › Documents › GLOSSARY.
Janić, M. 2021. System Analysis and Modelling in Air Transport: Demand, Capacity, Quality of Service, Economics, and Sustainability, CRS Press, Taylor & Francis Group.
Kanafani, A.K. 1983. Transportation Demand Analysis, McGraw-Hill.
Littlewood, K. 1972. Forecasting and Control of Passenger Bookings. AGIFORS Symposium Proc. 12, Nathanya, Israel.
Morrell, S.P. 2007. Airline Finance, Ashgate Publication Co, United Kingdom, 3rd ed.
Morrell, P. 2013. Airline Finance. Farnham, Surrey, England: Ashgate Publishing Limited and Burlington, USA: Ashgate Publishing Company, 4th ed.

Pitfield, D.E., Caves, R.E. and Quddus, M.A. 2010. Airline strategies for aircraft size and airline frequency with changing demand and competition: a simultaneous-equations approach for traffic on the North Atlantic. Journal of Air Transport Management 16(3): 151–158.

Shaw, S. 2004. Airline Marketing and Management, Fifth edition, Ashgate.

Shearman, P. 1992. Air Transport: Strategic Issues in Planning and Development, Pitman Publishing.

Tretheway, M.W. and Oum, T.H. 1992. Airline Economics: Foundations for Strategy and Policy, CTS, University of British Columbia, Vancouver, Canada.

Vasigh, B., Fleming, K. and Tacker, T. 2018. Introduction to Air Transport Economics: From Theory to Applications, Routledge, 3rd ed.

Wei, W. and Hansen, M. 2005. Impact of aircraft size and seat availability on airlines' demand and market share in Duopoly Markets. Transportation Research Part E: Logistics and Transportation Review 41(4): 315–327.

Wei, W. 2006. Impact of landing fees on airlines' choice of aircraft size and service frequency in Duopoly Markets. Journal of Air Transport Management 12(6): 288–292.

Wei, W. and Hansen, M. 2007. Airlines' competition in aircraft size and service frequency in Duopoly Markets. Transportation Research Part E: Logistics and Transportation Review 43(4): 409–424.

Wensveen, J. 2015. Air Transportation: A Management Perspective. Routledge, 8th ed.

Wittmer, A. and Hinnen, G. 2017. Airline passengers. Chapter 9. *In*: Budd, L. and Ison, S., Routledge (eds.). Air Transport Management: An International Perspective. Taylor & Francis Group, London and New York.

Chapter 7
Air Cargo

The transportation of cargo is profitable to the transport industry. Rail, water, and road cargo transport provide greater profitability than passenger transport. The only exception is air transport, which is thought to be intended for passenger transport primarily. What is evident is that the air cargo industry has been recording rapid growth in the last decades with an average rate higher than the average passenger transport growth rate per year.

During the regulation period, air cargo was unpopular in the airline industry due to high costs and uncompetitive transport tariffs. In the past, airlines used aircraft for transport of passengers in the cabin, while cargo was carried in the aircraft belly as supplement to the full payload, enabling additional earnings to be generated for airlines. Nowadays, things have changed, and certain airlines have opted for air cargo exclusively, thus becoming highly profitable.

Air cargo transport (airfreight) has proven to be a very important part of both the national and global economy. It has shown a constant growth trend since 1970, wherein the quantity of goods transported by air (expressed in RTKs) has doubled every 10 years. Air cargo represents less than one percent of global trade by tonnage, but at the same time, more than 35 percent of global trade by value (Boeing 2018). According to a forecast made by Boeing (2018), in the next two decades percentage of the average annual growth of air cargo, expressed in RTKs, will be 4.2, with a possible increase from 3.7–4.7%.

7.1 The history of air cargo

Transport of cargo by air started in 1910 in the USA. Two interesting events occurred during that time—a sack of mail was transported from Albany to New York City for the Post Office Department, and five bolts of silk were transported by air from Dayton to Columbus, Ohio. According to Wensveen (2015), there are three different elements of air cargo services, viz.:

- Airmail—transport of mail by aircraft,
- Air express—it used to include air freight, but today refers to small packages that usually have a higher priority of carriage than air freight, and
- Air freight—a term that appeared with the introduction of all-cargo aircraft.

Each of these elements has its own history that can be traced.

Airmail was developed earlier than other elements of air cargo services and it was a very important factor in creating air transport in the USA. In the 1920s, a significant increase in the volume of airmail has been recorded. Moreover, airmail was the main source of income until the aircraft DC-3 was introduced. This aircraft shifted the attention from the transport of mail to passenger transport, and from day to day, the transport of passengers became more profitable. The share of income induced by airmail transport has been reducing from year to year; therefore, today percentage of the airline's revenue induced by airmail is less than three (Wensveen 2015). The reason for reduction of airmail share could be found in the highly developing technology and shifting from ordinary mail to emails.

Air express, unlike airmail, did not have a significant impact and share in the early stages of development. Air express has started to meet expectations in recent history. This element of air cargo services was introduced in 1927 when some airlines joined the Railway Express Agency (REA) to provide freight service. Some other airlines did not want to join REA since it was competing for land transport, and in 1932 they established General Air Express serving 125 cities in the USA by direct routes. In the early 1930s, the revenue of air express represented 1% of total airline revenue. In 1973, Frederick Smith established Federal Express services at 13 airports offering overnight transport of small packages, from door to door. He also introduced the hub-and-spoke system where all small packages had their routes through the hub in Memphis, where students worked overnight, sorting and loading packages into aircraft. In the third year after its foundation, Federal Express gained a large income, but it was still losing money. It became profitable when the competing union of REA and the airlines broke up. From that period, this company has been steadily growing, and today it services one half of the total volume of small packages in the USA (Wensveen 2015).

According to the primary definition, airfreight represents the transport of goods by cargo aircraft, i.e., aircraft specialized for transport of freight only. This service appeared for the first time in the market in 1931 when Transcontinental and Western Air performed transport from New York to Kansas City, during the night, but it was not the scheduled transport. Cargo flights were performed when the capacity of the aircraft was completely used. The first scheduled cargo transport started in 1940 on the New York–Chicago route, operated by United Airlines (Wensveen 2015).

7.2 Air cargo today

The introduction of a cargo version of jumbo jet led to the establishment of cargo airlines that are dedicated to freight transport only. Examples of these airlines are Cargo Lux, Nippon Cargo, etc. Nowadays, we can distinguish different types of cargo airlines. According to Morrell and Klein (2019), these are:

- Passenger airlines,
- Combination carriers,
- Aircraft, Crew, Maintenance, and Insurance (ACMI) providers,
- Scheduled all-cargo carriers,
- Charter cargo carriers,
- Integrators.

Passenger airlines operate passenger aircraft only, while cargo is carried in the aircraft belly as a supplement to the full payload, enabling additional earnings for an airline. Examples of these airlines are Delta, American Airlines, British Airways, Swiss, etc. Other types of passenger airlines are limited to cargo transport due to their business models. Regional airlines operate small capacity aircraft; therefore, they do not have space for additional cargo transport. On the other hand, low-cost airlines tend to simplify all their processes. Thus, some of them (Ryanair, Wizz Air, EasyJet) do not accept cargo, while others accept it only while operating certain aircraft types (Norwegian accepts cargo carriage on its short-haul and long-haul flights operated by Boeing 737–800 and Boeing 787).

Combination carriers have both passenger and all-cargo aircraft in their fleet; therefore, they offer capacity for the transport of passengers and goods. Examples of these airlines are Emirates, Lufthansa, Air France-KLM, Cathay Pacific, etc. It is worth noting that some of them have a cargo fleet integrated into the passenger fleet, while others have their subsidiary (Lufthansa Cargo) dedicated to freight transport. However, the basic capacity for cargo transport is aircraft belly. Bearing in mind that cargo transport is the coproduct of passenger transport, prices for cargo are lower in comparison to integrators' prices. The income generated from cargo is about 15% of the total income.

ACMI (Aircraft, Crew, Maintenance, and Insurance) providers or wet lease means that a carrier operates cargo or passenger aircraft on behalf of another airline. An example is Atlas Air.

Scheduled all-cargo carriers are airlines whose fleets consist of cargo aircraft only. Examples of these airlines are Cargolux or Nippon Cargo.

Charter cargo carriers differ from scheduled ones because they do not operate scheduled transport. It means that they are used for ad-hoc situations, military necessity, or in a different crisis. An example of this type of carrier is Volga-Dnepr.

Integrators or integrated carriers offer door-to-door services. Their network is hub-and-spoke. They have cargo aircraft in their fleets but also have trucks, delivery vehicles, and advanced information systems. Examples of these operators are FedEx, UPS, or DHL.

Table 7.1 shows leading cargo airlines ranked according to the scheduled international freight tonne kilometres (FTK) in 2017 and 2018, (IATA 2018; 2019). Seven out of ten leading airlines in 2017 and 2018 are combination carriers, two of them are integrators and only one is an all-cargo carrier.

As above mentioned, air cargo is a developing part of the airline industry, characterized by uni-directional flows. In consideration of the IATA 2019, it can be seen that the largest number of freight tonnes carried in 2018 originated from the People's Republic of China. The second originating country was the United States of America, followed by Hong Kong (China). The top ten originating countries according to freight tonnes carried in 2018 are given in Table 7.2.

When it comes to freight country pairs at the world level (Table 7.3), the largest freight tonnes carried can be observed from the People's Republic of China to the United States (642,145), from Hong Kong to the United States (392,219), from the United Kingdom to the United States (387,864), from Japan to the United States (349,358) and from the People's Republic of China to Japan (322,861). It also can be observed that the largest freight flows go to the United States, which is not surprising because freight goes from markets with cheaper labour to the developed markets. Moreover, uni-directional flow can be seen in the example of the United States and the People's Republic of China, where the flow from China is almost three times larger than flow from the United States to China.

With the development of air cargo transport, the network of airports also started to develop. New cargo terminals that represent independent technology units are being opened. These terminals receive, store and ship the goods. Furthermore, in these terminals the goods are handled, which

Table 7.1. Scheduled international FTK in 2017 and 2018 for top 10 international cargo airlines.

Rank	Airline	2018	2017	Type
1	Emirates	12,713	12,715	Combination carrier
2	Qatar Airways	12,695	10,999	Combination carrier
3	Cathay Pacific Airways	11,284	10,772	Combination carrier
4	Federal Express	8,455	8,040	Integrator
5	Korean Air	7,815	7,988	Combination carrier
6	Lufthansa	7,391	7,313	Combination carrier
7	Cargolux	7,322	7,322	All-cargo carrier
8	Singapore Airlines	6,491	6,592	Combination carrier
9	United Parcel Service	6,252	5,894	Integrator
10	China Airlines	5,912	5,741	Combination carrier

Source: IATA WATS 2018; 2019

Table 7.2. Leading originating countries in 2018.

Rank	From	Freight tonnes carried
1	People's Republic of China	3,254,447
2	United States	3,077,624
3	Hong Kong	2,486,162
4	United Arab Emirates	1,795,717
5	Japan	1,691,811
6	Republic of Korea	1,439,666
7	Germany	1,437,683
8	Qatar	1,121,475
9	United Kingdom	1,100,082
10	India	1,080,969

Data source: IATA WATS 2019

includes all operations before loading and after unloading cargo from the aircraft (preparation for customs clearance, customs clearance) (Wensveen 2015). Terminals are equipped with special mechanisation and they might have a different level of automation of acceptance, transhipment, transport, preparation, storage, and shipping processes. In recent years, special information systems have been developed aiming to monitor the movement of cargo, as well as to issue and process appropriate documentation.

Table 7.3. Leading freight flows in 2018.

Rank	From	To	Freight tonnes carried
1	People's Republic of China	United States	642,145
2	Hong Kong	United States	392,219
3	United Kingdom	United States	387,864
4	Japan	United States	349,358
5	People's Republic of China	Japan	322,861
6	Republic of Korea	United States	298,521
7	United States	People's Republic of China	286,017
8	People's Republic of China	Republic of Korea	276,037
9	Germany	United States	272,145
10	United States	United Kingdom	263,234

Data source: IATA WATS 2019

Despite the steady increase, air cargo transport still does not represent a major source of revenue for airlines, although it is expected to do so shortly because in all other modes of transport freight transport brings more money than passenger transport. The question is why this is the case. The answer to this question lies in the fact that air transport is the most expensive mode of transport, compared to land or water transport. Hence, the customer must have a very compelling reason to use airfreight services. Examples of regular users are IT companies. For them, the most important concern is to have their product on a specific market as soon as possible, therefore, transport costs are negligible. Another example is related to the transport of perishable goods, such as fresh fruits, flowers, fish, etc., which has no alternative in other modes of transport. The second reason is that aircraft are constructed mainly for carrying passengers (Wensveen 2015). They have been converted into cargo versions by changing the passenger doors to cargo doors on the side of the aircraft (left), but also by installing the tail or nose doors. Therefore, the fact that passenger transport by air is more profitable than air cargo transport is straightforward reasonable.

Considering circumstances on the market caused by the Coronavirus pandemic (2020–2021), it could be noticed that the demand decline has been the steepest drop since 2009, during the global financial crisis. The share of bellyhold capacity in total air cargo capacity before the pandemic was almost 2/3. Due to the pandemic and significant decrease of passenger transport, as well as due to conversion of passenger aircraft to freighters or utilization of passenger aircraft as freighters, the bellyhold capacity share declined to less than one third of total air cargo capacity. However, cargo operations helped airlines survive the hard time for the aviation industry by providing some income while passenger transport was halted. The arising question relates to the bellyhold capacity share after the pandemic of Coronavirus—would it be the same as before or the situation in the market would change?

7.3 Air cargo fleet

The airfreight in the form existing today started to develop after World War II. Depending on the aircraft used for cargo transport in the fleet, the development of air cargo services could be divided into three periods (Dožić and Babić 2015). The first period encompasses the period after World War II, when there was a lot of unused transport military aircraft. These military aircraft had piston plant, short range, small speeds, small capacity (measured in mass unit), and small volume. To use their spare capacity, these aircraft were employed to carry valuable shipments, such as gems, precious metals, small packages, and similar things.

The second period in airfreight development started in the late 1950s with the appearance of aircraft with jet power plants (Dožić and Babić 2015). These aircraft are called narrow bodies and they have higher speed, larger range, and larger payload. They are used for the transport of passengers in the cabin, while cargo is carried in their belly as a supplement to the full payload. Hereby, an airline can generate additional earnings. The narrow body aircraft offered a possibility to transport different types of cargo, while the transport cost per unit of cargo was reduced compared to the unit cost in the previous period. The cargo door dimensions and belly volume, as well as the volume and mass of cargo consignment, limit the type of cargo that could be transported. Nevertheless, depending on the cross-section of the narrow body hulls, different cargo units (pallets and/or containers) can be used, which brings numerous disadvantages related to the fact that many different units of cargo require a great diversity of handling equipment, and that storage capacity is used inadequately.

The third period of airfreight history began in the 1960s with the advent of wide body aircraft that provided the opportunity to transport a variety of goods by aircraft (Dožić and Babić 2015). Initially, cargo was shipped in the bellies of passenger wide body aircraft as a supplement to the full payload. After that, with the development of air cargo, the aircraft that carried cargo in both belly and main deck were produced and allowed even unusual shipments to be transported. These aircraft are known as freighters or cargo aircraft and they offer possibilities to transport even up to 250 tonnes of goods on the same flight. The first freighter was delivered to German airline Lufthansa in 1972, and it was Boeing 747, the so-called jumbo jet (Wensveen 2015). This aircraft operated on the Frankfurt-New York route. It should be mentioned that mechanical and powered cargo loading systems are provided in freighters, which enables better stacking and securing of cargo, and reduction of cargo loading and unloading times, as well. These improvements enable further reduction of transport unit costs. The ability to transport a large number of containers and pallets of different sizes, as well as transport of intermodal containers, is provided by wide body freighters. Having in mind that air cargo transport surely includes intermodality because the cargo must be moved to and from the airport by a surface mode of transport (rail or road), intermodal containers are very suitable. Besides, the high capacity of freighters is convenient for the transport of oversized and heavy cargo (e.g., helicopters, aircraft fuselages, trains, etc.).

Apart from passenger aircraft and freighters, wide body aircraft are produced in combi and quick change version, as well. The main feature of combi aircraft is that both passengers and cargo are carried on the main deck of combi aircraft. It means that passengers may occupy the front section of the aircraft, with cargo occupying the back section. Sabena World Airways introduced the first combi Boeing 747 into scheduled service (Wensveen 2015). On the other hand, quick change aircraft are used as passenger aircraft by day and as freighters by night. They are equipped with quick change systems that include passenger seats on flooring panels. The panels are put on and off the aircraft like cargo containers.

About freighter aircraft, it could be said that they can be divided into different categories. As stated by Boeing (2014), freighter aircraft are divided into three categories and these are large wide-body, medium wide-body, and standard-body freighters. Airbus (2014) also distinguished three categories of freighters, viz.: large freighters, mid-size freighters, and small freighters. What is worth noting is that each freighter category corresponds to a particular market.

According to Boeing (2014), the share of wide-body freighters in the airline cargo fleet is about 65%, corresponding to 95% of the total cargo fleet's capacity. Bearing in mind the range limitation of small and mid-size freighters, it is not surprising that large freighters are usually employed on long-haul routes with high market demand. On the other hand, mid-size freighters operate regional routes where usage of large freighters represents a capacity risk, and demand is not small. Finally, due to their versatility and flexibility, small freighters are used when the terrain is geographically difficult, for example in the mountains, forests, or islands.

It is worth noting that market conditions may have a strong influence on world freighter fleet growth. A global downturn causes a decline in cargo demand and cargo flows, as well as a decline of cargo aircraft demand. Airlines (or other aircraft operators) face the problem of capacity surplus, which further leads to parking a certain number of their aircraft. Different factors, influencing the market condition, often cause contrary effects (Boeing 2014). For example, high fuel prices increase air cargo transport costs, which further decreases demand for services. Because converted passenger aircraft constitute more than half of freighters, it is not surprising that high fuel prices and fuel costs are the reason to replace old, less efficient freighters with newer, more efficient ones. Thus, the contrary effect is that the demand for new freighters is induced when fuel price is high. On the other hand, the demand for freighters decreases when yields, load factor, and utilization are reduced. The demand for freighters could be slightly reduced by the advent of new wide-body passenger aircraft with large cargo compartments. However, freighters are inevitable because they offer a specific advantage that could not be offered by passenger aircraft. Finally, the competitive mode of transport and bellyhold capacity could also influence the demand for mid-size freighters. Namely, they reduce demand for air cargo service by offering lower prices.

7.4 Air cargo market

Clients that want to send their shipments by air need to consider the speed, reliability, and quality of air transport. Furthermore, depending on the goods that should be shipped, these attributes can be considered as main advantages and crucial factors when choosing the mode of transportation. Clients must select the optimal transport mode and the optimal carrier that will deliver the goods to the final destination and to customers (Hsu et al. 2009). The charge, flight frequency, and transit time of the carrier, including the product characteristics (high pricing and/or perishable characteristics lead to a high inventory loss), influence the logistics cost of the shipper. Shippers forwarding products of high value on short distances consider the shipping charge as an influential factor and they are interested in high frequency services. Due to the competition with other modes of transport, freighter operators are sensitive to tonne-kilometre costs. To increase the market share, freighter operators should also adjust their supply attributes to the product characteristics (Hsu et al. 2009).

The air transport is most convenient for cargo (Wensveen 2015), when:

- Goods have at least one of the following characteristics:
  - Perishable,
  - Economically perishable (obsolete),
  - Required in the short period (express shipments),
  - Valuable,
  - Handling or storage is expensive;

- Demand is:
  - Unpredictable,
  - Infrequent,
  - Greater than local supply,
  - Seasonal;
- Distribution problems involve:
  - Risk of steal, breakage, or deterioration,
  - High insurance costs for long in-transit periods,
  - Heavy/complicated or expensive packaging required for surface transport,
  - Need for special handling or care,
  - Warehousing or stocks in excess of what would be needed if airfreight were used.

The speed of air transport is a crucial factor when perishable, economically perishable or urgently needed goods are being shipped. Examples of these types of goods are CDs, DVDs, high fashion apparel, newspapers, etc. On the other hand, when the seasonal market or when the demand changes for some reason, air cargo provides a quick response without inventory costs. For producers having a wide range of products and a geographically wide selling network, air cargo offers a possibility to provide products for the market from a central warehouse at the same speed as from a local warehouse.

Commodities that have been transported by air are usually high value and high quality commodities. As mentioned before, the share of air cargo is less than one percent of the global trade by tonnage, but it accounts for more than 35% of global trade by value (Boeing 2018). It is not unusual that commodity transported by air obtains an added value, and makes the producer more convenient in the eyes of customers. Air cargo transport provides an option for producers to enter a new market without having additional investments related to warehousing and inventory. Furthermore, deliveries for the market are done overnight as in the case of the local market.

The next important feature of air transport refers to the risk of stealing, breakage, or deterioration, which is minimised because the commodity is better secured and fewer people can be in contact with the shipment. Insurance charges are lower for air freight compared to surface freight because shorter transit time leads to lower risks.

The cost of packaging goods and preparing shipments for air cargo transport is usually lower than that of the surface modes of transport. Usually, a cardboard carton package is sufficient for air transport; while heavy wooden crates may be necessary for surface transport. The exception is a special load that requires special conditions of transport (for example, medical materials usually require temperature-controlled air freight containers).

7.5 Air cargo demand characteristics

The basic factors influencing air cargo demand are as follows (Boeing 2014):

- The economic development and level of trade. The trend of air cargo transport is very cyclical and largely depends on the global economic conditions. The more favourable the economic conditions and the higher the trade exchanges (imports and exports), the higher the demand for the transport of goods by air.
- The price of air cargo transport compared to alternative modes of cargo transport, including water, road and rail cargo transport.
- The increase in demand of air cargo transport largely depends on the increase in passenger demand (primarily on long-haul routes). The increase in passenger demand influences the increase in capacity on a considered route, and thus increases the capacity intended for cargo transport on the given route (larger aircraft have larger cargo space).

In addition to the factors that stimulate the growth of air cargo demand, there are certain factors that can adversely affect air cargo transport, such as (Boeing 2014):

- *Economic recession*, which usually causes reduction in production, reduction in demand for certain goods, reduction in international trade, etc., which further results in air cargo demand decrease.
- *Barriers to trade*, such as customs barriers and different kinds of protection of domestic production that restrict free trade.
- *Regulations related to aircraft*, such as noise levels and emissions reduction, as well as increasing safety. For example, implementation of noise reduction equipment can reduce aircraft payload, and consequently the amount of paid cargo that can be transported.
- *Competition* of other modes of transport, especially in shorter distances, where air cargo transport is considerably more expensive compared to road or rail cargo transport.

Unlike passenger transport, air cargo transport has some specifics that need to be taken into account during the planning process. Airlines that transport goods usually face the problem that flows of goods on given routes are expressed only in one direction. This type of problem occurs especially on routes from and to countries that are engaged only in the production or import goods, and on routes that connect points with different levels of economic development. In this case, the airline must take these restrictions into account when designing the flight schedule. To avoid flying empty aircraft in one direction, airlines usually planned flights as a circular route, thus connecting the markets to have paid cargo on each flight.

Passenger airlines are obliged to provide capacity such that each passenger has their seat. For cargo airlines, the problem of providing capacity is not so simple. Namely, cargo is not homogeneous and shipments may vary from small packages lighter than 1 kg to parcels of 30,000 kg and more. Apart from the weight, the goods can also differ in density and easiness of storage (packaging). Moreover, cargo sent by air can be one of the very demanding services. Some goods require special equipment during loading and unloading. High value goods require special security; perishable goods require special conditions for storage and transport, etc. (more on special load will be provided in the following sections).

Differences between passenger and cargo transport on long-haul routes occur because passenger air transport has almost no competitive mode of transport on these routes. On the other hand, competition in cargo transport is present even on long-haul routes. The main competitors to cargo airlines on long-haul routes are rail and water transport, primarily due to lower transport costs. However, there are some routes where the prices are similar due to similar tariffs, but in most cases, the prices of air cargo transport are significantly higher compared to other modes of transport, sometimes reaching up to 10 times higher values.

7.6 Advantages of air cargo transport

Speaking of air cargo transport, certain advantages could be distinguished in comparison with other modes of transport (Doganis 2010):

- *Lower packaging costs*—due to more favourable conditions in which the goods are transported and less risk of damage, large/additional packaging is not necessary. Since less packaging material is used, the goods are lighter, and therefore the price of transport is lower.
- *Insurance costs* are lower due to more favourable transport conditions.
- Air cargo transport brings a significant *flow of money*—in the cargo transport, services are always charged upon delivery of the goods. In other modes of transport, the service provider may wait several weeks before charging for the service, while in air transport it usually takes one or two days.

7.7 Handling units in air cargo

A Unit Load Device (ULD) is either an aircraft pallet and pallet net combination, or an aircraft container (IATA 2020b). ULDs represent units into which or on which one could put shipments for consolidation. ULDs fit into the structural characteristics of the aircraft with their shape (cross-section of the fuselage), as Figure 7.1 shows, and the fastening system that is installed in the cargo compartments of the aircraft (either in the lower deck of passenger aircraft or in both the lower and the main deck of cargo aircraft). It should be noted that upon loading in the aircraft these units become an integral part of the aircraft, which is important because this prevents damage to aircraft structure by cargo movements. At the same time, ULDs secure the content therein against possible damage. ULDs are removable aircraft parts that are designed, tested, and produced according to strict civil aviation authorities' requirements. ULDs also need to be repaired and maintained following the requirements. A ULD must be structurally capable to hold the loads and provide adequate protection to the aircraft systems and structure during flight to be airworthy (IATA 2020b).

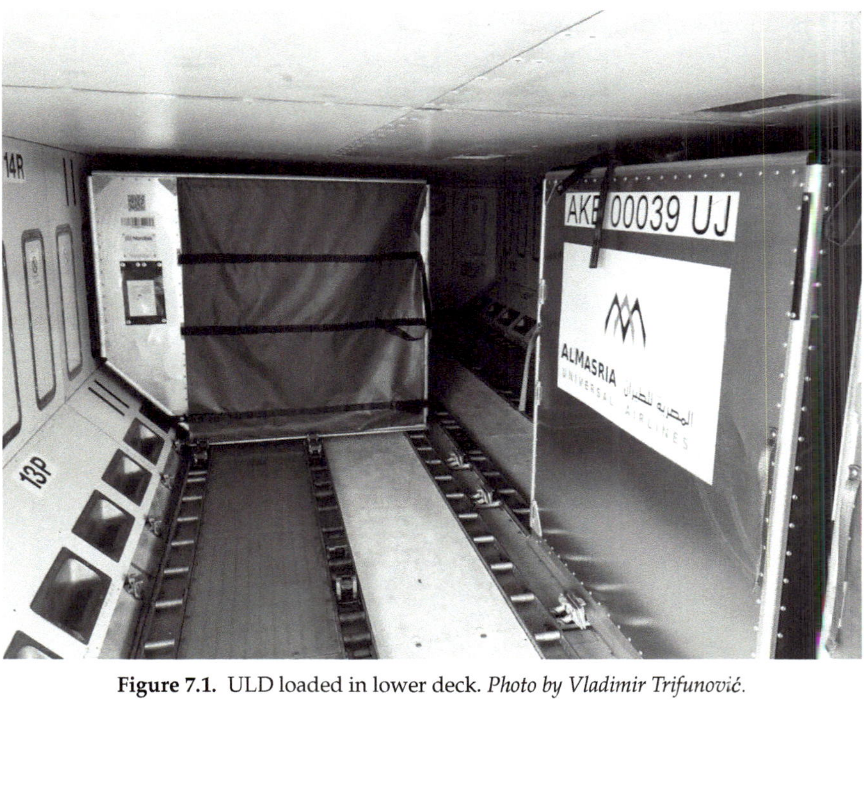

Figure 7.1. ULD loaded in lower deck. *Photo by Vladimir Trifunović.*

Using ULD ensures some advantages for an airline. Namely, it reduces the time needed for loading, and unloading processes, provides better protection against weather conditions and better protection against damage to the cargo or to the aircraft, as well. Furthermore, the personnel working in the built-up area could be less experienced as the contour/shape is complete, and fits in the aircraft structure. Pallets also have certain advantages because some cargo is difficult to fit into containers. Moreover, there are more combinations of shipments (options) for build-up when using an open pallet. Furthermore, some special loads can be loaded only on open pallets.

As mentioned before, cargo can be transported by passenger aircraft which can be narrow bodies (known as non-ULD) and wide bodies (known as ULD carrying aircraft), and by cargo-only aircraft (known as freighters). It is worth noting that most narrow bodies accept only loose cargo, which is sometimes referred to as bulk loads, and passenger baggage. However, some of these aircraft accept pallets and/or containers to fill in the lower deck, if there is empty space, after passenger baggage loading. Wide bodies can accommodate ULDs in the lower deck, but also have a loose load compartment called bulk load. Finally, freighters carry cargo in both the main deck (Figure 7.2) and the lower deck, thus offering a vast amount of cargo space.

Figure 7.2. Cargo loaded in main deck of freighter. *Photo by Vladimir Trifunović.*

7.8 Material handling and transport equipment

Material handling and transport equipment are divided into two groups, the continuous material handling equipment and material handling equipment with cycle time (Sretenović 1996). Continuous material handling equipment includes equipment for horizontal, oblique, and vertical transport for bulk and piece cargo (belt conveyors, roller conveyors, plate conveyors, conveyors, hanging conveyors, etc.). Material handling equipment with cycle time is used for internal transport, i.e., to move cargo. This equipment is divided into carriages which can be electro-carriages or auto-carriages, forklifts, elevator platforms that are hydraulically driven and which are used for handling heavy loads, aircraft high loader (Figure 7.3), air pallets and containers, tractors with trailers, various transport-handling vehicles without propulsion, car cranes, rack elevators-stackers, etc.

Note that all material handling and transport devices are equipped with rollers and ball bearings that facilitate the handling of ULDs and speed up the process of loading, reloading, and unloading.

Figure 7.3. Aircraft high loader. *Photo by Vladimir Trifunović.*

7.9 Air freight transport documents

Different stakeholders are involved in the air cargo transport process. Namely, the main stakeholders are usually the consignor, the consignee, the freight forwarders, and the airline. According to ICAO and WCO (2016), a consignor is an entity or an individual who initiates the movement or transport of the goods. Precisely, the consignor is the sender. Another term can also be found, and it is the "shipper" which is used to describe the entity or an individual who initiates the trade in goods. Although consignor and shipper can be the same entity or an individual, these are separate roles in air cargo supply chains. The recipient of goods at the end of transport is known as the consignee, and this party is designated on the invoice or the packing list (ICAO and WCO 2016). The main task of freight forwarders is to arrange for air shipments (which might include consolidation of cargo) to be managed in such a way that they are ready for transport by an airline that is responsible for the transport of goods. In addition to stakeholders mentioned above, ground handlers and airports could be responsible for and in contact with shipments in some phases of transport.

Ground handlers conclude subcontracts with freight forwarders and/or airlines; therefore, they act on behalf of them, usually when the freight forwarder and/or airlines do not have the necessary facilities. Ground handlers are responsible for dealing with operational aspects, which are in accordance with the instructions of freight forwarders and airlines. An airport is responsible for the provision and security of the airport infrastructure. It should provide a secure environment through which the goods move and may also be responsible for the provision of 'on-airport' cargo services (ICAO and WCO 2016).

The air cargo movement could be illustrated by the scheme presented in Figure 7.4. Freight forwarders pick up consignments/shipments from the consignor/shipper, and they consolidate the shipment, if necessary. Further, the shipment is forwarded to the ground handler at the origin airport who delivers it to the airline. After that, the airline transports the goods by air and, at the destination, the shipment is taken over by a ground handler again. The ground handler sends the shipment to the freight forwarder at the destination where the shipment is deconsolidated, if necessary. In the end, the freight forwarder delivers the shipment to the consignee.

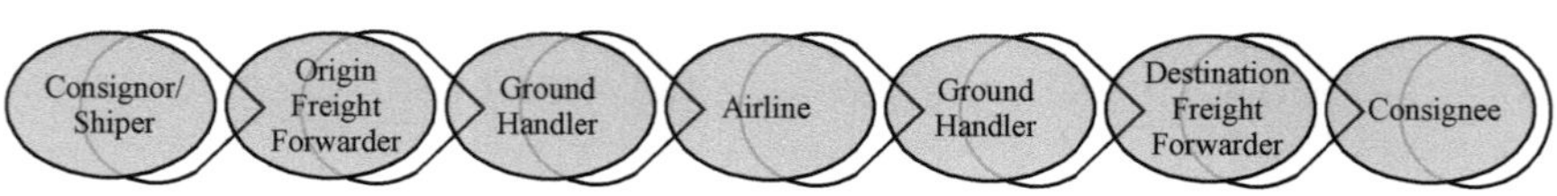

Figure 7.4. Air cargo movement.

Cargo shipped by aircraft needs to be accompanied with appropriate air freight transport documents. One of the basic documents in air cargo transport is the Air Waybill (AWB). The AWB is a document prepared by or on behalf of the shipper that evidences the contract between the shipper and the airline(s) for the carriage of goods over routes of the airline(s) (ICAO and WCO 2016). It is a non-negotiable document that indicates possession of a shipment and which serves as a receipt for the consignor. The AWB can be issued by an airline either directly or through its authorized agent (freight forwarder) and covers the transport of cargo from airport to airport. The AWB used to be issued in hardcopy form in three originals and eleven copies. One original was dedicated to the airline, the second was for the consignee and the third one was for the consignor, while one of eleven copies served as a Delivery Receipt. In the year 2010, the Electronic Air Waybill (e-AWB) was introduced, and nine years after its introduction, the e-AWB became the default contract of carriage for all air cargo shipments on enabled trade lanes on 1 January 2019. Due to The Electronic Air Waybill Resolution 672 (MeA), there is no longer a need to print, handle or archive the paper AWB. Therefore, by removing the requirement related to paper AWB, the air cargo process is largely simplified (IATA 2020a).

Air Cargo Manifest is a document that contains the details of consignments loaded on to a specified flight and provides a list of all the air waybill and master air waybill numbers referring to the goods loaded onto the aircraft. This document contains information such as the nature of the goods, weight, and the number of pieces composing each consignment on a specified flight, and the unit of loading used, as well. The airline issues the air cargo manifest and it could be found in hard copy or electronic form.

Air Cargo Transfer Manifest is a document that defines conditions of transhipment. By signing the Air Cargo Transfer Manifest, the responsibility is transferred from the previous to the next airline.

The other required documents include the Shipper's letter of instruction (SLI). It is a document that the consignor provides to the freight forwarder and that states the characteristics and requirements of the cargo, the contact person for any questions concerning the cargo or the export itself, and the reason for the export.

When accepting documents and shipments, the airline should check the accuracy of the data, namely the number of parcels (packages), shipment weight (due to maximum load capacity, and determination of the appropriate tariff), shipment dimensions (due to cargo door size, cargo loading, and determination of appropriate tariffs), the contents of the consignment (appropriate tariff, packaging, type of consignment), packaging of the consignment and the correct use of labels and markings on the parcels.

Upon acceptance of documents and shipment, the airline acknowledges receipt of the shipment and the AWB in cases where an AWB is issued and completed by an authorized IATA cargo agent or the airline takes over the AWB and cargo from another airline. The airline issues an AWB based on SLI in case it accepts the shipment directly from the shipper and when it accepts the shipment from a non-IATA cargo agent.

7.10 Special loads

The term "special loads" refers to those items of cargo that, due to their value or their features, require special treatment during the stages of acceptance, storage, stowage, loading/unloading, and transport, because they are sensitive to changes in temperature, pressure, humidity, and lighting. Special care is needed to protect the aircraft, crew and passengers, the personnel who come into contact with such shipments, other shipments, mail, and luggage, as well as the special load (self-destruction) from their possible harmful effects.

Special loads transport requires specific terms of transport. These terms and methods of transport are defined by ICAO and IATA regulations, as well as regulations issued by the state of airline registration and airline's Ground Operations Manual (GOM). Airline's and state's regulations need to be in accordance with the regulations issued by ICAO and IATA, while airline's regulations need to be in accordance with the regulations issued by the state. The purpose of GOM is to provide detailed guidance for procedures and functions for aircraft ground handling, equipment, operational procedures, training programs, and organizational structure. These manuals are divided into chapters dedicated to administration, safety, staff training, standard service agreements and contracts, cargo apron, managing of cargo handling units, handling of dangerous goods, passenger service, cargo handling, equipment, quality, etc. The manuals should ensure that all activities of the handling agent are performed in accordance with airline procedures, user requirements, but also in accordance with national regulations and international standards. GOMs are updated to be in line with changes in the applicable international standards and regulations. It should be noted that special loads can only be handled by persons who have been trained to do so and who have the appropriate licenses.

For the transport of any special load, it is necessary to inform the captain via the Notification to Captain (NOTOC) form about the contents of that particular shipment. The NOTOC form must be completed and the information it contains must be presented to the captain before the start of the flight. Shipments must be loaded in accordance with the compatibility list which defines types of goods that may be in the same trunk, the distance that is needed between them, whether another cargo is needed in between, etc. Cargo space reservations must be made in advance. Packaging must be adequate and in accordance with the existing regulations. An IATA document, The Air Cargo Tariff and Rules (TACT), provides instructions for handling, storage and transport of goods, regulations of all countries, airport codes, airport capacities and suitable equipment (warehouses and cargo handling equipment), as well as tariffs for transport of goods. The sender guarantees to the airline that the shipment is in appropriate condition, and the airline is obliged to verify it itself when feasible (e.g., there is an available veterinarian, phytopathologist, etc.). The shipment must be accompanied by appropriate, prescribed documentation, namely export, import, passport (for animals), death certificate, veterinary certificates, etc. The shipment should bear the address of the recipient, but also the address of the sender, as well as their telephone numbers. It is also necessary to have required labels and marks. Before shipping the shipment, it is necessary to check whether the airport(s) through which the goods are supposed to go have the appropriate equipment, storage space and licensed personnel who can handle the shipment. The airports of transfer (if any) and the destination should be notified of the type of consignment arriving, while the airport of destination should inform the airport of departure that the consignment has arrived and has been forwarded to the consignee.

An airline provides in the GOM types of special loads. It should be noted that these types of special loads are classified differently, depending on the airline. The special loads can be generally classified as:

- Perishable cargo,
- Valuable cargo,
- Heavy cargo,
- Human remains,
- Express shipments,
- Service cargo,
- Diplomatic cargo,
- Live animals,
- Dangerous goods.

Each class of goods will be described in more details in the next sections.

7.10.1 Perishable cargo

The term perishable cargo refers to goods whose characteristics or suitability for their original or prime purpose may deteriorate below their useable condition or may become completely unusable if exposed to inappropriate changes in temperature and humidity or delayed in transport. They can be divided into three groups: foodstuffs (EAT), perishables that are not foodstuff (PER) and vaccines, blood and blood plasma, and live human organs (LHO).

Foodstuffs are shipments of plant and animal origin that can be used for human consumption, and they include fresh fruits and vegetables, meat and fish shipments. These shipments can also be classified as wet shipments due to the presence of water, and their transport requires waterproof packaging. Perishable shipments that are not foodstuff include flowers and shipments that are economically perishable (newspapers, films, exhibits, etc.), i.e., those shipments that lose their economic value if they are not delivered to the destination at a certain time. Due to the presence of water, shipments of vaccines, transplant organs, and blood plasma are also classified as wet shipments, and their transport requires waterproof packaging, as well.

To transport perishable cargo successfully it is necessary that a shipment be packed in an appropriate manner and in accordance with the applicable regulations, loaded into compartments in accordance with the compatibility list (the list which defines the types of goods that can be loaded in the same compartment) and accompanied by appropriate documents. For example, perishable shipments should not be loaded in the same compartment with poisons, live animals, or human remains. Furthermore, it should be known that film strips are not transported in the same trunks with radioactive materials and magnets. Flowers should be transported packed in cardboard boxes lined with paper or cellophane, covered with paper and a lid. These boxes should be placed as far away from the sides and floor of the compartment (because these places are the coldest) and as far away from the fruit as possible due to ethylene (fruit releases ethylene, which causes flowers to wither).

The Perishable Cargo Regulations (PCR) manual issued by IATA (2021c) is the leading publication for temperature control and cold chain management of goods and it provides information on the latest efficient practices for adequate shipping of time and temperature sensitive goods. Fruits and vegetables are transported in cardboard boxes with openings (due to ventilation), at a temperature of 7–10°C, where the boxes are stacked on top of each other, and there is a suitable space between them, which is necessary for ventilation. When fresh meat is transported, it is necessary to disinfect the compartment (before and after transport). The meat is wrapped in gauze for hygienic reasons. Cooling is performed with dry ice (CO_2 in solid state, at a temperature of –80°C) placed in boxes with openings to prevent an explosion. The temperature in the trunk is maintained at 0–5°C, and if frozen meat is transported, the temperature is –12°C, PCR (IATA 2021c).

Transporting vaccines, blood and medication quickly and safely become more important than ever during the Coronavirus pandemic (2020–2021). Due to the fact that these products are time and temperature sensitive, and that the people they are meant to help often don't have time to wait, IATA (2021d) issued the Temperature Control Regulations (TCR) which contains all needed information.

7.10.2 Valuable cargo

Each airline defines in its GOM what can be considered valuable cargo (VAL). They also define conditions for transport, as well as packaging. Shipments of valuable cargo are usually stowed in valuable-cargo pouches with attached seal, and with the defined dimensions and weight of the shipment. To enable secure transport, special handling procedures are required for valuable cargo, during acceptance and loading into the aircraft compartment. Thus, they are loaded last in the bulk compartment and unloaded first.

7.10.3 Heavy cargo

If the individual item weighs 150 kg or more,[1] it is considered Heavy cargo (HEA). The GOM defines the number of heavy items that can be loaded on the compartment for each aircraft type. Their maximum dimensions and masses are also defined. These dimensions and weights are important due to the dimensions of the cargo door and the payload that is allowed for the aircraft, as well as due to floor bearing capacity. It should be noted that although their weight may be greater than 150 kg, human remains are not considered as heavy cargo. After loading, heavy cargo should be fastened in an appropriate, reliable manner, to prevent possible movement during the flight. It is also necessary to check that the transfer airports (if there is a transfer) and destinations have the equipment needed to unload these shipments.

7.10.4 Human remains

Human remains (HUM) can be transported as non-cremated and cremated, and they differ in the way of packaging. Non-cremated human remains shall be contained in a metal container that is hermetically sealed. That

---

1 https://www.iata.org/contentassets/6c6b8373246b4b2db532ff9c89bee5a7/iata-shc-occ-compliance-monitoring-tool-v1.0.xlsx

container is placed in a coffin, and then in another neutral wooden coffin, on the basis of which it cannot be inferred that the remains are being transported therein. Due to the specific packaging (metal and two wooden containers), the mass of the remains can be over 150 kg. In addition to the non-cremated remains, it is necessary to submit appropriate documents, namely the certificate of death, passport and addresses (of the sender and the recipient) shall be attached to the cargo manifest. It is also necessary to mark these shipments with appropriate labels and marks that indicate the correct way of handling them. When loading or unloading, non-cremated mortal remains are transported (to/from the aircraft) in a trolley covered with a tarpaulin. Other goods that according to the compatibility list can be in the same compartment must be at a minimum distance of 1 m. Under no circumstances human remains shall be loaded into the same hold/ULD as live animals or perishables. The shipment with non-cremated human remains should spend as little time as possible on the cargo platform, and be stored in the refrigerator in the warehouse if it is needed.

Cremated remains are transported in an urn in a sealed outer box. Styrofoam is placed between the urn and the box. Such shipments must be loaded into the belly of a passenger aircraft (their transport in the passenger cabin is not allowed).

7.10.5 Express shipments

Express shipments are those shipments that need to be delivered to the destination in a short period. Some airlines deal exclusively with the transport of express shipments, for example, DHL, TNT, FedEx, UPS. Express shipments include transport of documents, letters, small packages, but also larger packages (Worldwide Parcel Express—WPX) which are transported door to door. They are loaded into the bulk compartment.

In addition to the listed shipments, service cargo is also included in express shipments. This type of shipment has absolute priority. It is transported for use by the airline or for another airline according to the existing contract. These shipments include transport of spare parts, catering food, catering containers, advertising materials, etc. An air waybill is issued for each consignment, and the contents of the consignment are entered.

7.10.6 Live animals

Transport of live animals by air is very specific. The Live Animal Regulations (LAR) publication issued by IATA (2021b) once a year defines all conditions necessary for transport of live animals. Each edition contains certain changes compared to the previous one, and those who participate in the transport of live animals must be aware of these changes. The LAR need to be in accordance with the CITES (*Convention on International Trade in Endangered Species of Wild Fauna and Flora*), OIE (*Office International des Epizooties*) and authorities of different countries.

The European Union has adopted LAR as the minimum necessary conditions for the transport of live animals in containers, boxes, and pens. Consignors of live animals must fully comply with these rules, as well as any additional rules laid down by national authorities in the country of origin, transit, or destination.

Since live animals are transported, it is necessary to take into account their behaviour. They are instinctively afraid of the unknown environment which occurs during the transport realization. To minimize possible difficulties, special containers are used during the transport of animals. Moreover, these containers are adequately designed to keep the animal, in accordance with all reference regulations. Some of the specific recommendations refer to the fact that transport of mammalian pups is not recommended, that animal feeding must be taken into account during transport, as well as the reaction of a stressed animal. Transport conditions depend on the animal species and they should be complied with. Thus, for example, some animals should be transported in the dark, while others should be transported with adequate lighting.

Regulations related to primary factors should be met in the animal's surroundings. The primary factors are temperature, humidity, air flow, and concentration of ammonia and carbon dioxide. The adaptability of an animal to temperature changes varies depending on its age, mental state, whether the animal is fed or not, as well as whether it has had to suffer temperature changes recently. Consignments with live animals should be placed in the compartments so that the change in temperature caused by cargo door opening is not sudden. Different species of animals can be transported in different, previously defined temperature ranges. The temperature is controlled during transport by the aircraft heating system or at the airport by ground vehicles for air conditioning, or using blankets and various insulating materials.

Humidity should also be within a defined range. High temperatures and high humidity interfere with the release of heat through sweating and increase the risk of an animal's temperature shock. High humidity at low temperatures increases heat release and can cause shock due to cooling. Humidity must be controlled by adequate ventilation, and if necessary, the possibility of temperature adjustments during transport should be considered.

For animals, the air flow can be fatal, because it accelerates heat loss. Therefore, animals should not be exposed to the air flow influence. To protect the animals from wind, rain, snow, or extreme temperatures, the containers can be partially covered, but only for a short time to prevent suffocation. Animals must not be exposed to heat directly. When the temperature change exceeds 20°C, the cargo door must be open for some time and in extreme temperature conditions, ground equipment must be used to maintain adequate conditions in the compartment. Changes in climate factors that may appear during animal transport must be forecasted and considered while determining the route.

When it comes to the conditions for the transport of live animals, it should be noted that depending on the animal species being transported, there are appropriate, prescribed conditions to be adhered to. Thus, for example, wild animals are mostly transported in the dark or semi-darkness, because they can rest in such an environment. Domestic animals and most birds prefer to travel in low light because it calms them down. Animals that are natural enemies shall not be placed in the same compartment. Quarantined animals must be isolated from others. Animals intended for laboratory use must not be kept close to other animals to reduce the risk of infection. Shipments containing primates from different continents must be isolated from each other in compartments, warehouses, animal storage rooms, as well as during all phases of transport.

The responsibility of the sender in the process of transporting live animals implies that one should:

- emphasize the sex of the mammal being transported,
- obtain all necessary documents and correct information for shipper's certification,
- act in accordance with the airline's, national and international regulations (LAR),
- provide containers that comply with the latest edition of the LAR,
- provide the necessary substrate in the container and food,
- provide the scientific and trivial (common) name of the animal species and the number of individuals in containers,
- emphasize the need to feed and water the animal,
- in a visible place on the container, emphasize the instructions, the date and time when the animal was last fed and watered,
- emphasize if the animal was pregnant or had a litter in the past 24 hours,
- emphasize if the animal is under medical treatment: medicine, dose, time when the medicine is given, etc.

While preparing an animal for transport by air, consignors need to have all relevant information related to export/import, necessary permits in transit countries, veterinary certificates, for endangered species—a certificate that export is permitted and a copy of the document that import is permitted in the destination country, veterinary examination, quarantine, data on prohibitions, if any. These rules are subject to frequent changes and depend on the species of animals, and are published in the LAR. The sender must provide a telephone number on which someone will be available 24 hours a day, so that the airline can receive any necessary information if it is needed, as well as in case of emergency. This telephone number must be indicated on the air waybill.

When it comes to airlines, it should be noted that there are several airlines that do not accept live animals for transport. The only exceptions are guide dogs for the blind. Also, airlines that transport live animals do not accept all species for transport, or accept some of them but only under special conditions. The airline that realizes the transport must, in addition to the transport itself, make a reservation of the cargo space, as well as ensure the delivery of the animal.

As the airline is responsible for the reservation of cargo space, it is common for the reservation of space to be made before the animals are handed over for transport to the airline. When planning the transport, the route should be used with as few landings and take-offs as possible, to avoid changes related to the handling of the shipment, as well as changes of the climate. It should also be taken into account that sometimes certain countries ban the import of animals that have passed through areas where there was an epidemic of an animal disease in the period immediately before. Therefore, it should not be flown over airports located in such countries. Since some airports do not have available customs and veterinarians during holidays and weekends, the animals should not arrive at the airport during these days. It should be noted that departure and arrival times at the destination airport defined in the flight schedule are approximate and they are not a part of the transport contract. The airline is able to change the aircraft, the route, without the obligation to emphasize it in the air waybill. Also, airlines may limit the number of individual species that are present on the airplane. Furthermore, factors such as cargo door dimensions or luggage compartment size determine whether an animal can be accepted for transport in an animal transport container.

If there is a need to transfer the consignment, it should be noted that when several airlines participate in the transport, the animals are not accepted for transport unless each segment of transport is arranged in advance and confirmed by all participants. The first airline in the row responsible for transport must emphasize to each following airline the number of the air waybill; the number of individual shipments; weight and dimensions of the package; number and species of animals; departure, transit and destination airport; date of transport; name, address and telephone number of the recipient; as well as necessary mechanization for cargo handling.

Upon completion of the transport, it is necessary to ensure the delivery of the animal to the recipient as soon as possible from the moment of arrival at the destination airport. The sender is obliged to inform the recipient about the planned arrival of the animal. To successfully realize the delivery immediately upon arrival at the destination airport, the customs authorities should be informed about consignments containing live animals. If a consignment of live animals cannot be delivered to the consignee, the airline should behave in such a way that provides the shipment is undamaged as well as to inform the first airline and the consignee of all actions taken.

Appropriate documents are required for transport of live animals, as follows: Shipper's Certification for Live Animals, Air waybill (AWB), Notification To Captain (NOTOC), CITES (Convention on International Trade in Endangered Species of Wild Flora and Fauna) documents, animal passport with the veterinary certificate, and Live Animals Acceptance Check List.

The Shipper's certification for live animals is intended to confirm that the consignment is properly packaged, marked, in accordance with IATA LAR regulations and regulations issued by the government of a country or the airline, to show that the consignment meets the requirements of the national legislation and that the consignor accepts that the airline will not be liable for any loss, damage or expense resulting from natural disasters, force majeure, or the consequence of the animal injuring itself. This certificate should indicate the number of containers required for the transport of animals, the species, scientific or trivial (common) name and the number of units, as well as the airport of departure and destinations, date, and consignment note number.

A separate air waybill is filled in for live animals (they should not be on the same air waybill with other types of goods). In the section where the type and quantity of goods should be stated on the air waybill, the class of the animal should be entered in English. The number of individuals must be indicated in the air waybill, as well as the telephone number of the person who will be available at all times.

The pilot-in-command must be informed which species are onboard the flight, as well as where they are located and how many of them are in the aircraft. Everything needs to be in accordance with the IATA document Airport Handling Manual Procedures. The NOTOC form, with information on loading-related details and necessary actions, must be presented to the captain before the flight starts.

CITES (Convention on International Trade in Endangered Species of Wild Flora and Fauna) documents must be issued for each individual, as follows: export permission, import permission, re-export permission, certificate of origin, a certificate that the animal is part of an exhibition group and confirmation that an exchange is taking place between scientific institutes or scientists.

The consignor is required to provide health certificates and permissions prescribed by national authorities as mandatory. It is prescribed that the animal must have a passport that contains all relevant data related to the individual being sent, as well as a confirmation that the animal has been vaccinated, if it is sent to European Union countries or if a transit station is in a European Union country. After vaccination, it takes a month for the transport of animals to be allowed.

To facilitate the process of accepting animals, both for senders and agents and airlines, the Live Animals Acceptance Check List has been standardized. To further verify the shipment, it is advisable to ensure that national regulations are met during the inspection. When filling out this list, each question must be answered with "YES" or "NO". If there is at least one "NO" answer, the shipment does not meet IATA regulations for shipments of live animals and it shall not be accepted for transport.

Containers/pallets for the transport of live animals should also comply with the relevant regulations.[2] Thus, animals are transported in closed containers, while transport in open pens must be specially arranged. The container must be adequately constructed. The dimensions should be in accordance with the size of the animal for which it was constructed. Ventilation and other needs related to a certain species need to be satisfied as well. An appropriate container is dedicated for each animal species, which is further in accordance with the needs of the animal (defined in the LAR). For example, for certain species, it is necessary to use reinforced containers (due to the size and weight of the animal), while for others it is necessary to use metal containers (due to the possibility of animal's destructive behaviour). Also, it is recommended that the container for live animals' transport should be easy to handle. Moreover, special requirements need to be satisfied to provide secure transport of animals. For example, the spacers must be in line with the design, to allow the movement of the container and to prevent blocking the ventilation system by other goods. The container must be constructed in such a way as to prevent the handling personnel from being bitten or injured by the animal. The construction of the container must be such that it is able to withstand damage possibly caused by other goods, such as breakage of the structure or dents, but it also must prevent the animal from getting out of it. The joints of wooden containers must be made so that the animal

---

2 https://www.iata.org/contentassets/b0016da92c86449f850fe9560827bbea/pet-container-requirements_1_47e_en.pdf

cannot damage them from the inside by scratching or milling. The interior of the container needs to provide safe transport of animals, without the possibility of injury for the animals. This means that the edges must not be sharp, but smooth and rounded, and the walls need to be without sharp protrusions. The size of ventilation openings should have such dimensions to prevent any part of the animal body from being out of the container, and yet large enough to provide adequate ventilation. The animal should be able to stand, turn and lie in a natural way in the container. The exceptions are bird cages in which there must be enough perches for each individual and enough space for the birds to stand on, so that their tail does not touch the floor of the container. Birds that do not stand on their perches should have enough space to stand upright. To reuse the container, it is necessary to clean it thoroughly and then disinfect or sterilise it before reuse. The container must be clean and leak-proof. The shipper is obliged to provide absorbent bedding that absorbs the liquid, and which corresponds to the type of animal being transported. The container must be constructed of non-toxic materials.

Bearing in mind that live animals are transported, it is necessary to provide enough water and food for them, due to the possibility of a prolonged journey. Separate food and water troughs must be fixed inside the container or attached to it so that they are accessible for replenishment purposes. Their construction needs to have rounded edges and be made of non-toxic material suitable for the species. The shipper's instructions related to watering and feeding the animals should be submitted in written form, together with the animal. The shipper is obliged to provide food, but in doing so one must check whether the rules of the countries of transit and destination are violated. Products of animal origin, such as meat or food containing meat, must not be placed into the container.

Every consignment of live animals must be labelled in accordance with the LAR instructions. The shipper is responsible for labelling. Labels must not block ventilation openings, especially when labelling small containers. All labels and instructions must be written in the English language. The label must contain information about the sender; the scientific and trivial name of the animal and the number of individuals of each species contained in the container (in accordance with the shipper's certificate of live animals); for specific animals, an appropriate label must be affixed (e.g., a container containing venomous animals must be labelled "Venomous", for aggressive animals or birds that may cause injury through grilles or vents "This animal bites", etc.); instructions related to feeding and watering the animal. Sedatives are given to some animals during transport; therefore, in that case, it is necessary to clearly indicate the name of the sedative, the time of falling asleep, and the route on the container.

7.10.7 Dangerous goods

Dangerous goods in air transport are articles or substances whose chemical and/or physical properties may pose a danger to passengers, crew, aircraft, and everything in the environment, and are in the list of dangerous goods, IATA (2021a) Dangerous Goods Regulations (DGR), or are classified according to it (definition according to DGR).

The basic documents related to the transport of dangerous goods are ANNEX-18 of the Chicago Convention (The Safe Transport of Dangerous Goods by Air) and ICAO Technical Instructions for the Safe Transport of Dangerous Goods by Air. These documents contain detailed instructions for the safe transport of dangerous goods by aircraft. The regulations given in the IATA edition of the Dangerous Goods Regulations are fully compliant with ANNEX 18, or stricter. States and airlines whose regulations differ from ICAO and IATA regulations are required to notify ICAO and IATA of such deviations, and these deviations are set out in the ICAO Technical Instructions for The Safe Transport of Dangerous Goods by Air, as well as in IATA Dangerous Goods Regulations.

There are several categories of dangerous goods that can be accepted for transport:

- Dangerous goods that are acceptable for transport under the conditions defined by the IATA DGR,
- Dangerous substances that are prohibited for air transport,
- Dangerous substances with hidden danger,
- Hazardous substances exempted under IATA DGR,
- Hazardous substances that are acceptable for air transport under certain conditions.

A vast majority of dangerous goods can be transported by aircraft as a consignment that is properly prepared for transport in accordance with IATA DGR. Such dangerous substances are not allowed to be transported in the luggage of passengers or crew, and they are not allowed to be brought into the aircraft as personal belongings.

Dangerous substances prohibited for air transport include: explosives that are flammable or decompose if they stand for 48 hours at a temperature of 75°C, explosives containing both chlorine and ammonium salts, explosives containing mixtures of chlorine salts and phosphorus, solid explosives in particular susceptible to mechanical shock, any substance, offered for carriage, capable of producing dangerous heat or gas under normal conditions of air transport, radioactive liquids which are flammable and combustible liquids and organic peroxides possessing explosive properties and which are packaged in such a way to require a label for explosives as well as a label for the additional risk.

Dangerous substances with a hidden danger include items that may pose a danger that is not obvious (such as camping equipment, refrigerator, dental appliances, some instruments (thermometers, barometers containing mercury, etc.), items that can be found in passenger luggage (flammable household liquids, corrosive oven cleaners or dry cleaners, flammable gas or lighter fillers, matches, bleach, and the like).

For the following items, the sender should confirm with the Material Safety Data Sheet that they do not contain hazardous substances:

- cars, car parts—may contain ferromagnets, batteries, etc.,
- breathing apparatus—may contain bottles of compressed air or oxygen,
- camping equipment—may contain flammable gases, flammable liquids, matches,
- samples for diagnosis—may contain infectious substances,
- electrical equipment—may contain magnets, mercury in switches and electronic tubes,
- electrically powered appliances—may contain batteries (wheelchairs),
- frozen fruits, vegetables, etc.—can be packed with dry ice.

Exempted hazardous substances are aircraft parts, equipment, and supplies needed in the passenger cabin (alcoholic beverages, perfumes for sale in the aircraft, dry ice for cooling food and beverages during the summer), as well as containers specially designed for the transport of aircraft parts (aircraft batteries, without weight limit, oxygen bottles, etc.).

Exempted hazardous substances can be found in both passenger luggage and personal belongings. Alcoholic beverages are allowed in checked luggage and personal belongings of passengers or crew, up to a maximum of 5 l (net amount) in one container. Medical or toilet articles, whose total net quantity per passenger or crew member does not exceed 2 kg or 2 l, or when the net quantity of an individual item does not exceed 0.5 kg or 0.5 l are allowed.[3] This category of dangerous goods includes items such as hairspray, perfumes, and varnishes that contain alcohol. Ammunition for sports purposes is also excluded, if it is safely packed, and the maximum allowed amount of packages must not exceed 5 kg per passenger which should be placed only in checked luggage with the approval of the airline. It should be noted that explosive and flammable ammunition is not allowed to be transported in luggage. Dry ice belongs to the category of exempted hazardous substances if it is transported in

[3] https://www.iata.org/contentassets/6fea26dd84d24b26a7a1fd5788561d6e/dgr-62-en-2.3a.pdf

the amount of up to 2 kg per passenger and if it is used to store perishable items, but only in personal belongings. Matches and lighters are also allowed in personal belongings while charging for lighters is not allowed. Mercury barometers may be transported in the personal belongings of meteorological service representatives and need to be approved by the airline. They must be packed in a steady outer package, while the inner package must be sealed. It is also possible to use a bag made of a mercury impermeable material as the inner packaging. The captain of the aircraft must be provided with information related to such a barometer on the flight. Self-propelled wheelchairs can also be transported with the approval of the airline. They must be packed according to packing instructions, IATA DGR (IATA 2021a), as checked baggage.

In addition to the listed hazardous/dangerous substances, the airline itself defines certain hazardous substances, as well. If the hazardous substances of the airline are articles and substances that are integral parts of the aircraft equipment, the regulations do not apply to them. Examples of such equipment are fire extinguishers, oxygen cylinders, oxygen generators, rescue equipment, first aid kits, dry ice used to cool food and beverages, Duty Free (sprays, alcoholic beverages, perfumes, etc.), safety matches, liquid lighters gas, etc.

Dangerous Goods Classification. According to the United Nations (UN), we can distinguish nine hazard classes of dangerous goods, and regardless of how cargo is being transported, it must comply with these nine hazard classes. There are rules for each class that determine how its transport is performed. Moreover, there may be additional instructions for each substance separately. Many substances have such properties that they can be classified into several groups, so the rules for all these groups apply to them at the same time (for example, ammonia is a gas and therefore belongs to group 2, but it is also toxic, so the rules of group 6 should be applied to it, as well).

The nine hazard classes for dangerous goods are following:[4]

Class 1 – Explosive materials,

Class 2 – Gases,

Class 3 – Flammable Liquids,

Class 4 – Flammable Solids,

Class 5 – Oxidizing Substances and Organic Pesticides,

Class 6 – Toxic and Infectious Substances,

[4] https://unece.org/DAM/trans/danger/publi/unrec/rev17/English/02EREv17_Part2.pdf

Class 7 – Radioactive Materials,

Class 8 – Corrosive Materials,

Class 9 – Miscellaneous.

Class 1 encompasses explosive materials and it is not usually shipped by air. This class is divided into 6 subdivisions that cover substances that are explosion hazards, explosions that may project fragments and firebrands, and fire hazards.

Class 2 is related to gases and it is divided into three subdivisions. These subdivisions include flammable gases, toxic gases, and gases that are neither flammable nor toxic, such as helium and oxygen.

Class 3 is related to flammable liquids and comprises liquids or mixtures of liquids that will give off flammable vapours at specific temperatures and have a flash point of not more than 60.5°C/140.9 F.

Class 4 encompasses flammable solids. The class is divided into 3 subdivisions that include highly flammable solids, solids that are likely to spontaneously emit flammable gases, and substances that, if they come into contact with water, emit flammable gases.

Class 5 comprises oxidising substances and organic pesticides, and it is divided into two subdivisions. Subdivisions cover agents that react with oxygen and organic pesticides.

Class 6 is related to toxic and infectious substances. It is divided into two subdivisions and includes substances such as cyanide, arsenic, vaccines, and pathology specimens.

Class 7 covers radioactive materials that have a specific activity greater than 70 kBq/kg.

Class 8 covers corrosive materials and does not have any subdivisions. The class comprises corrosive liquids and solids that will cause severe damage when in contact with living tissue; or, of leakage, will materially damage or even destroy other goods or the aircraft itself. Corrosive items encompass battery acids, sulfuric acid, and mercury.

Class 9 is dedicated to miscellaneous dangerous items. The class does not have any subdivisions but comprises any substance that may pose a danger during air transport that is not covered by the other classes. This includes items with anaesthetic properties, solid dry ice, asbestos, life rafts, and chain saws.

Labels and Marks. It should be noted that each class and subdivision have appropriate labels and marks that provide information necessary for the safe handling of such shipments. Packaging of dangerous goods must comply with the IATA DGR, and the airline is required to verify that the shipment is properly packaged. There are two basic types of packaging. The first one is a combined packaging, which means that the articles or materials are placed in inner packages, and this package is placed in one or more outer packages. The second type of packaging is single packaging, where dangerous goods are placed in one container (barrels made of plastic, steel, or aluminium). Usually, such packages are lined or coated with some other material. They can also be complex packages, which consist of an outer and an inner package but are constructed in the shape of a single package that cannot be separated into parts.

Dangerous goods should be labelled and marked appropriately following the regulation. There are two types of labels, and they are:

- hazard labels which are square-shaped and where the side of the square is set at an angle of 45° to the horizontal line,
- handling labels and marks which are rectangular and that could mark dangerous goods by themselves or together with the hazard labels.

All labels should have appropriate dimensions and must be printed in English. For example, the minimum dimension of a hazard label must be 100 mm × 100 mm. In addition to the inscription in English, there may be an inscription in the language of the relevant country (consignor, consignee, or transit country). The labels must indicate the class of dangerous goods being transported. The size of letters and numbers must be at least 13 mm.

When it comes to appropriate documentation, it is necessary to provide a statement from the consignor about dangerous goods. This document ensures that the consignor takes responsibility for acting per the regulations, as well as for any possible damage caused to the airline if the damage is not the result of improper handling of dangerous goods by airline staff during transport.

The responsibility of the consignor also includes checking whether the goods/substances are prohibited for air transport, identification of the substance, classification of the substance, packaging in appropriate, approved packaging, labelling and marking the package, preparation of documents.

The airline is responsible for acceptance, storage, and loading packages, as well as for checking for possible damage, providing and forwarding information to all necessary participants, establishing procedures for emergencies and staff training, as well as storing data related to training.

References

Airbus. 2014. GMF 2014 Freight: Flying on Demand. http://www.airbus.com/company/market/forecast/(Accessed: March 2, 2015).

Boeing. 2014. World Air Cargo Forecast 2014–2015. http://www.boeing.com/assets/pdf/commercial/cargo/wacf.pdf. (Accessed: March 2, 2015).

Boeing. 2018. World Air Cargo Forecast 2018–2037. https://www.boeing.com/resources/boeingdotcom/commercial/about-our-market/cargo-market-detail-wacf/download-report/assets/pdfs/2018_WACF.pdf. (Accessed: 01.09.2020).

Doganis, R. 2010. Flying off course: airline economics and marketing, Routledge, 4th ed.

Dožić, S. and Babić, D. 2015. Trends in airline cargo fleet. pp. 172–177. *In:* Vidović, M., Kilibarda, M., Zečević, S., Miljuš, M. and Radivojević, G. Proceedings of 2nd Logistic International Conference, Belgrade, Serbia, University of Belgrade, Faculty of Transport and Traffic Engineering.

Hsu, C.I., Li, H.C., Liao, P. and Hansen, M.M. 2009. Responses of air cargo carriers to industrial changes. Journal of Air Transport Management 15(6): 330–336.

https://unece.org/DAM/trans/danger/publi/unrec/rev17/English/02EREv17_Part2.pdf (Accessed: 17.08.2021).

https://www.iata.org/contentassets/6c6b8373246b4b2db532ff9c89bee5a7/iata-shc-occ-compliance-monitoring-tool-v1.0.xlsx (Accessed: 17.08.2021).

https://www.iata.org/contentassets/6fea26dd84d24b26a7a1fd5788561d6e/dgr-62-en-2.3a.pdf (Accessed: 17.08.2021).

https://www.iata.org/contentassets/b0016da92c86449f850fe9560827bbea/pet-container-requirements_1_47e_en.pdf (Accessed: 17.08.2021).

International Air Transport Association (IATA). 2018. World Air Transport Statistics, Montreal – Geneva.

International Air Transport Association (IATA). 2019. World Air Transport Statistics, Montreal – Geneva.

International Air Transport Association (IATA). 2020a. e-AWB. https://www.iata.org/en/programs/cargo/e/eawb/(Accessed: 01.09.2020).

International Air Transport Association (IATA). 2020b. Unit Load Devicecs. https://www.iata.org/en/programs/cargo/unit-load-devices/. (Accessed: 28.04.2020).

International Air Transport Association (IATA). 2021a. Dangerous Goods Regulations (DGR), 62nd ed.

International Air Transport Association (IATA). 2021b. Live Animals Regulations (LAR), 47th ed.

International Air Transport Association (IATA). 2021c. Perishable Cargo Regulations, 20th ed.

International Air Transport Association (IATA). 2021d. Temperature Control Regulations (TCR), 9th ed.

International Civil Aviation Organization (ICAO) and World Customs Organization (WCO). 2016. Moving Air Cargo Globally: Air Cargo and Mail Secure Supply Chain and Facilitation Guidelines. https://www.icao.int/Security/aircargo/Moving%20Air%20Cargo%20Globally/ICAO_WCO_Moving_Air_Cargo_en.pdf (Accessed: 28.04.2020).

Morrell, P. and Klein, T. 2019. Moving Boxes by Air: the Economics of International Air Cargo, 2nd ed. Routladge, Taylor and Frencis Group, London and New York.

Sretenović, M. 1996. Mehanizacija Pretovara: Pretovarne Mašine I Projektovanje Pretovarnih Procesa. Univerzitet u Beogradu.

Wensveen, J. 2015. Air Transportation: A Management Perspective. Routledge, 8th ed.

Index

B

C

F

G

H

I

J

K

L

M

N

O

P

Q

R

S